Introduction to Mechanisms of Hormone Action

Introduction to Mechanisms of Hormone Action

P. Catherine Clegg
BSc PhD MB ChB

Sometime Lecturer in Physiology, University of Sheffield

William Heinemann Medical Books Ltd
London

First published 1969

SBN 433 06210 0

Printed in Great Britain by The Whitefriars Press Ltd
London and Tonbridge

CONTENTS

PREFACE

The advent of modern cellular and molecular studies in Biology has considerably widened the horizons of endocrinology. At last we can speculate on the nature of the action of hormones on cells. This progress has been achieved mainly since 1961 which means that many elementary standard texts do not cover this new extension of endocrinology. It is hoped that this small volume will serve to introduce the subject to Upper School and University students.

Examples have been chosen to illustrate the diverse mechanisms by which hormones may modify cellular activities. It has been necessary to consider some basic aspects of cell structure and function to make the discussions comprehensible, and to give perspective I have also briefly discussed the roles of the various hormones in the whole organism.

Catherine Clegg.
Brighton, January 1969.

CHAPTER 1

Hormones, Their Nature and Sources

Introduction

In the early 17th century William Harvey published his studies which described for the first time the circulation of the blood. This discovery was critical to the development of physiological thought and without it a proper understanding of the processes of respiration, absorption, excretion, and so on, would have been impossible. It was, however, over two hundred years later that biologists became aware that in this circulating blood chemical substances were transported from special organs or tissues which produced them so as to exert profound effects upon distant tissues.

One of the earlier hints of this function of the circulating blood was provided by the observations of Berthold (1849). The effects of removing the testes from cockerels have been known for centuries. When this operation is carried out the comb and wattles of the cock shrink in size and lose their turgidity and brilliant colouring. The characteristic male plumage is lost and aggressive behaviour disappears. Such an animal is called a capon. In 1849 Berthold described experiments in which he transplanted testes from normal cocks into capons. The capons then redeveloped to a degree the characteristic features of the cock bird. These results suggested that some active substance was liberated from the transplanted testis which reached the tissues of the capon by way of the circulating blood. It was found later that testicular extracts could produce these same effects in the capon. In 1927 McGee isolated a relatively pure chemical substance from the testes of bulls which could produce similar masculinizing effects when injected into animals. The pure crystalline hormone was later synthesized.

In the late 19th and early 20th centuries a great variety of organ and tissue extracts were found to be capable of producing marked physiological effects when injected into animals. They have been found in a wide range of species of both invertebrate (insects, crustaceans, molluscs, etc.) and vertebrate (fishes, amphibia, reptiles, birds and mammals) types. The active principles present in these extracts have been given the name hormones.

Today there are literally hundreds of chemical substances, isolated from various tissues and organs, which can produce distinct effects upon growth, secretions, behaviour, etc., when they are injected into the bloodstream.

Local and General Hormones

In 1935 Huxley made the first attempt at the classification of the bewildering array of active substances isolated from animal tissues. He divided these substances, or activators into two broad groups:

1. Local Activators
2. Distance Activators:
   (a) diffusion activators
   (b) circulating activators

A vast number of chemical substances are synthesized within living cells. Many of these substances are confined to the cells producing them and these are included in the group of *local activators*. In certain cells some of the substances produced by the cell ultimately leave the cell (i.e. they are secreted) and some may have effects upon other tissues. These substances may reach their "target cells" by diffusion and thus their sphere of influence may be rather restricted. Huxley called these *diffusion activators*. For example the substances acetylcholine or noradrenaline are stored in the axon terminals of motor nerve cells. When such a nerve is stimulated then a small quantity of the stored substance (acetylcholine or noradrenaline) is released and diffuses across a microscopic gap between the nerve terminal and the end organ (e.g. striped or unstriped muscle, gland). These substances are neurotransmitters and link electrical events in the nerve cells with activity in the end organ.

Another group of activators, called *circulating activators* are distributed throughout the body in the circulating blood. It is these circulating activators which have come to be regarded as "hormones". There is however no clear line of demarcation between the two classes of distance activators. Indeed some substances may belong to both groups. Although noradrenaline is a diffusion activator at the terminals of sympathetic nerve fibres on the end organs it is also a circulating activator because it is released in large quantities from the medulla of the adrenal gland and is distributed throughout the body in the blood stream.

Sources of Hormones

Hormones have been isolated from many sites in the vertebrate organism. Perhaps the most familiar are those highly organized structures functioning as secretory units called endocrine glands. The term *endocrine* describes the way in which the secretions of the gland are discharged directly into the blood stream. These glands do not have the duct system which characterizes *exocrine* glands, whose secretions are discharged by way of a duct to external or internal surfaces of the body. Typical examples of endorine glands include the pituitary, thyroid and adrenal glands. All endocrine glands characteristically have a rich blood supply which provides not only the raw materials for hormone synthesis but also the transport system for the hormones.

Hormones are also produced from various tissues which are not so discreetly organized as endocrine glands. Thus the hormone insulin is secreted by small groups of cells scattered throughout the pancreas, the bulk of which is concerned in the production of a digestive juice. The male sex hormone is produced by groups of cells scattered between the seminiferous tubules of the testis.

Although the precise cellular origin of very many hormones is known there are other hormones about which we have no precise information concerning the location of their synthesis and release.

The Chemical Nature of Hormones

The many hormones which have been chemically identified fall into one of several classes of chemical substances including proteins, polypeptides, derivatives of amino acids, lipids and sterols. The body has thus utilized many kinds of substances to use as "chemical messengers". No great purpose would be served in this account of how hormones act by a detailed consideration of the chemical nature of all hormones. Table 1 gives a brief classification of the chemical nature of some hormones.

A consideration of the mode of action of individual hormones will most certainly require a detailed knowledge of the structure of the hormone molecule, its steric configuration, the kind of chemical bonds in the molecule and the electrical charges that it carries. Not all of the hormone molecule may be essential for the ultimate interaction of the hormone with the "receptor substances" in the target cell. Some parts of the molecule may act as a transporting vehicle which may protect the molecule from chemical inactivation during the course of its long journey through the blood stream, across the capillary wall into the interstitial fluid which bathes the target cells. Yet another part of the molecule may be involved in guiding the hormone onto the receptive substance in or on the target cell so that the "trigger" part of the molecule can come into appropriate connection with the receptive substance.

A great deal is known about the chemical nature of many hormones and about the effects of modifying the structure of the molecule on the biological actions of the hormone. It would, however, be fair to say that this information has, in general, given us little insight into the way in which the hormone triggers off its physiological effects. One reason for this is the elusiveness of the "receptor substances" of cells. It is now fairly easy to obtain relatively large amounts of many hormones for analysis of chemical and physical properties and to permit chemical modification of the hormone molecule; a study of the biological effects of modified molecules gives us indirect evidence for the functions of various regions of the molecule. The study of the properties of the "receptive substance" of cells with which hormone molecules interact presents much greater problems. We can often only hazard a guess at the structure and properties of this material, inferring that it must have a complementary nature to the hormone molecule with which it interacts. We may have no information of its location in the cell, and even if we have this information it is not normally possible to extract it from the cell in amounts necessary for study of its properties.

There have been some claims for isolation of receptive substance from target cells (page 30). One feature of the union of a drug or hormone with the receptive substance of the cell is that the changes in shape and electrical charges on the receptor, which are induced by union with the hormone molecule, have effects upon adjacent molecules. This propagation of the effects of receptor occupation may be very important in determining the effects of the hormone. If receptors are studied in isolation from their context in the cell then these propagated effects of receptor occupation are not seen. Figure 1 illustrates an example of a propagated action. Only the active part of the hormone molecule is shown and this contains a disulphide ring structure (S–S). The hormone receptor is located on the cell membrane and its active centre is a sulphydryl group (SH). Adjacent to the receptor is a pore penetrating the cell membrane. The diameter of the pore is small and there is resistance to dilatation of the pore by virtue of disulphide bridges linking proteins in the membranes on each side of the pore. On occupation of the receptor by the hormone there is

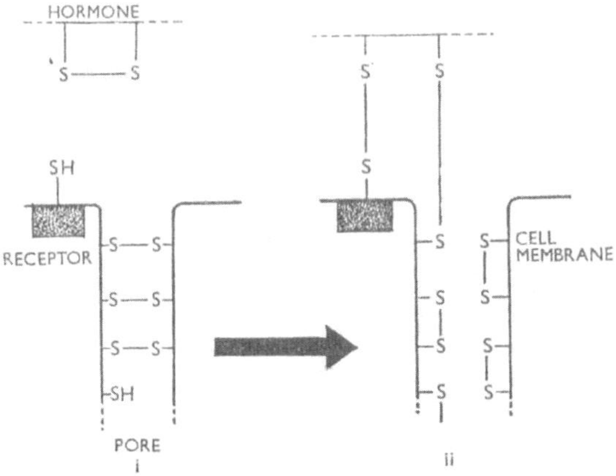

FIG. 1. *An example of a propagated effect of occupation of a receptor by an activator.*

Only the active chemical groups of the hormone and receptor molecule are shown. In (i) the hormone and receptor are not united. On occupation of the receptor (ii) there is a propagated action down the pore through the cell membrane resulting in widening of the pore.

---

an opening of the disulphide ring in the hormone molecule. This is followed by interaction of the hormone with the proteins lining the pore with fission of S–S links across the pore. This results in an increase in pore diameter and thus an increase in permeability of the membrane to solutes. If this receptor were isolated from its position on the cell membrane then receptor occupation would no longer be followed by changes in pore diameter and permeability.

An Example of Hormone Structure in Relation to Activity: Acetylcholine

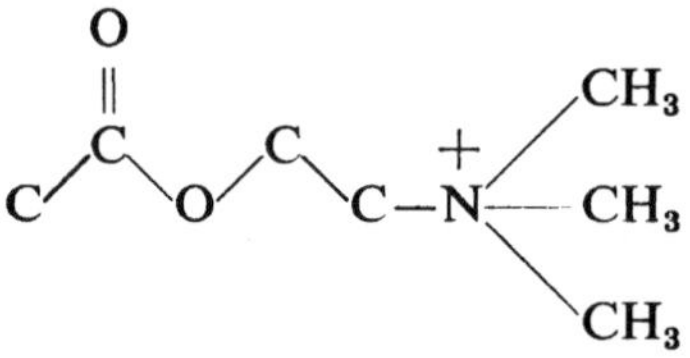

onium group

(1) The Importance of the Positively Charged Onium Group. If the molecule of acetylcholine is modified (see below) one finds a good correlation between the structure of the molecule and its biological activity. The positively charged onium group has been found to be very important in determining the activity of the compound. All the substances listed in Figure 2 can mimic to a degree the action

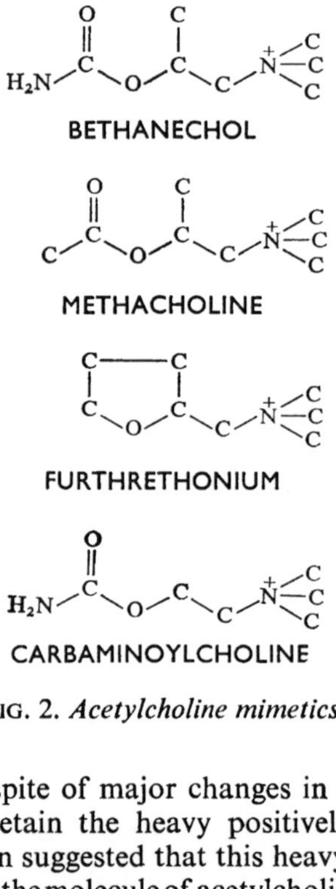

FIG. 2. *Acetylcholine mimetics.*

of acetylcholine in spite of major changes in the "backbone" of the molecule. All retain the heavy positively charged cationic grouping. It has been suggested that this heavy terminal charge is important in guiding the molecule of acetylcholine to its receptor site in the tissue. If the receptor substance has a complementary structure to the hormone then this we imagine bears a negative charge.

The biological activity of drugs which mimic acetylcholine is reduced if there is a reduction of the positive charge on the onium group, or if the approach of the hormone molecule to the receptor site is hindered either by substituting large groups on the onium group, or by substituents which shield the electrical field around the onium group. Thus the substitution of the larger ethyl group for methyl groups on the onion group reduces the activity. The ethyl groups hinder the approach of the drug to the receptor because of their larger size and because they shield the positive charge.

(2) The Chain in Acetylcholine. Although the positively charged onium group is of great importance in guiding the hormone to the receptor there are other and weaker forces involved. As the hormone approaches the receptor under the electrostatic force of the onium group, weaker forces become important in particular Van der Waal's forces and hydrogen bond formation. These come into action only as certain parts of the drug and receptor closely approach one another. The forces are important in binding the drug to the receptor. Thus, although the chain of the acetylcholine molecule can be dispensed with as long as the methylated onium group remains, very much higher doses of the modified compounds are required to produce the same effects as acetylcholine.

The effects of union of the positvely charged head of acetylcholine with the (presumably) negatively charged receptor is dealt with further in Chapter 3.

TABLE 1

Major chemical classes of Hormones

1. *Proteins and Polypeptides*	*Hormones*	*Major Sources*
Those containing fewer than 100 amino acids/ molecule are classed as polypeptides	Prolactin Follice stimulating hormone Interstitial cell stimulating hormone Thyroid stimulating hormone Adrenocorticotrophic hormone	Anterior Pituitary gland
	Oyxtocin (8 amino acids) Vasopressin (8 amino acids)	Posterior Pituitary
	Insulin (51 amino acids)	Islets of Langerhans

2. *Derivatives of Amino Acids*	*Hormones*	*Major Sources*
These derivatives, particularly decarboxylation products are substances of marked physiological activity and many hormones in this class have potent effects upon smooth muscle.	Noradrenaline (Tyrosine)	Chromaffine cells Sympathetic nerve terminals
The parent amino acid is shown in brackets	Adrenaline (Tyrosine)	Adrenal medulla
	Histamine (Histidine)	Widespread distribution
	Thyroxine (Tyrosine)	Thyroid gland
3. *Lipids*		
(a) Sterols (solid alcohols)	Cortisone Hydrocortisone Aldosterone	Adrenal cortex
	Oestradiol Progesterone	Ovary, Placenta
	Testosterone	Testis
(b) Fatty acids	Prostaglandins (polyunsaturated hydroxy-fatty acids)	Many tissues e.g. uterus, lung, brain, semen
4. *Quaternary Ammonium Cations*		
	Acetylcholine	Many nerve tissues. (page 18).

CHAPTER 2

The Multicellular Organization and Levels of Control

Multicellularity

The evolution of complex organisms seems to be dependent upon the body having many cells. Mammals, for example, attain a high degree of complexity of form and function and consist of many cells which are specialized to perform particular functions on behalf of the whole organism. Of course all body cells are similar in that they possess the basic "machinery" for the provision of energy, self-maintenance and the regulation of exchange of materials with the immediate environment of the cell. This basic similarity apart, mammalian cells show a high degree of differentiation of both structure and function. Nerve cells show to a degree the phenomenon of irritability and, when excited, they conduct messages in the form of a wave of electrical activity which surges along the long cytoplasmic extensions of the cell body of the neurone. The red blood cell is a highly specialized cell consisting almost entirely of the protein haemoglobin concentrated inside the enveloping cell membrane; it is a cell of no fixed site and lives its 120 days of life buffeted in the flowing blood engaged in the transport of oxygen from lungs to tissues and of carbon dioxide from tissues to the lungs. Muscle cells, like nerve cells, are excitable but on excitation the protein filaments of the muscle cell undergo changes in molecular relationship which results in shortening of the cell.

The Cell—Internal Control Mechanism

Multicellularity then allows the specialization of protoplasm for the more efficient performance of all necessary functions and is a prerequisite for the organization of the more complex animals. In each of the specialized cells lies a controlling system residing in the nucleus of the cell—i.e. the genetic material. As we shall see later the cell nucleus determines the structure and activities of a particular cell by controlling the kind and amount of proteins produced by the cell, and in particular those "catalysts" of biochemical activity called enzymes. However, in addition to this internal regulator the cell is subjected to external regulating influences which serve to link the activities of one group of cells with those of another, and also to adjust cellular activities to changes in the external environment (Fig. 3).

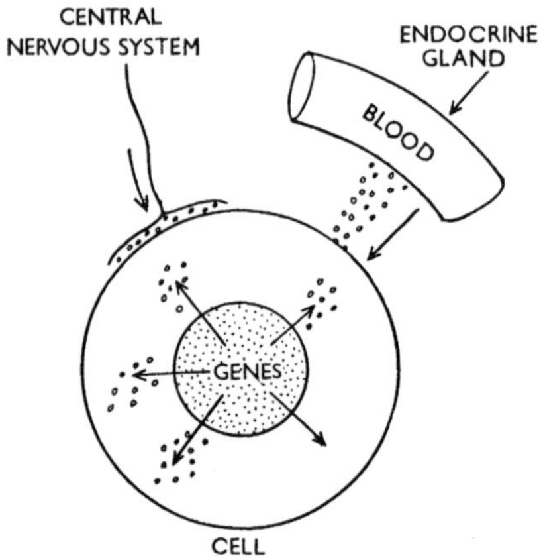

FIG. 3. *Regulators of cell activities.*

Two classes of external regulators are shown. Messages reach the cell from controlling centres of the body mainly in the form of nervous (i.e. electrical) or chemical messages (hormones). Note however that the last step in the transmission of nerve influences is in fact a chemical message (a neurotransmitter substance, e.g. acetylcholine, noradrenaline).

The internal regulator (genetic material) influences cell activities by releasing varying amounts of messenger substances into the cytoplasm of the cell.

External Control of the Cell

Nervous and Endocrine Mechanisms. The external control of cell activities is carried out by means of nervous and endocrine mechanisms. It is ultimately via the nervous system, either acting directly on cells or via the intermediary of endocrine secretions, that the individual cells of the body are provided with a link with the external environment. These two regulating mechanisms, nervous and endocrine, are closely linked to one another. In the floor of the forebrain there is an intimate link between the hypothalamus—that part of the brain concerned with basic

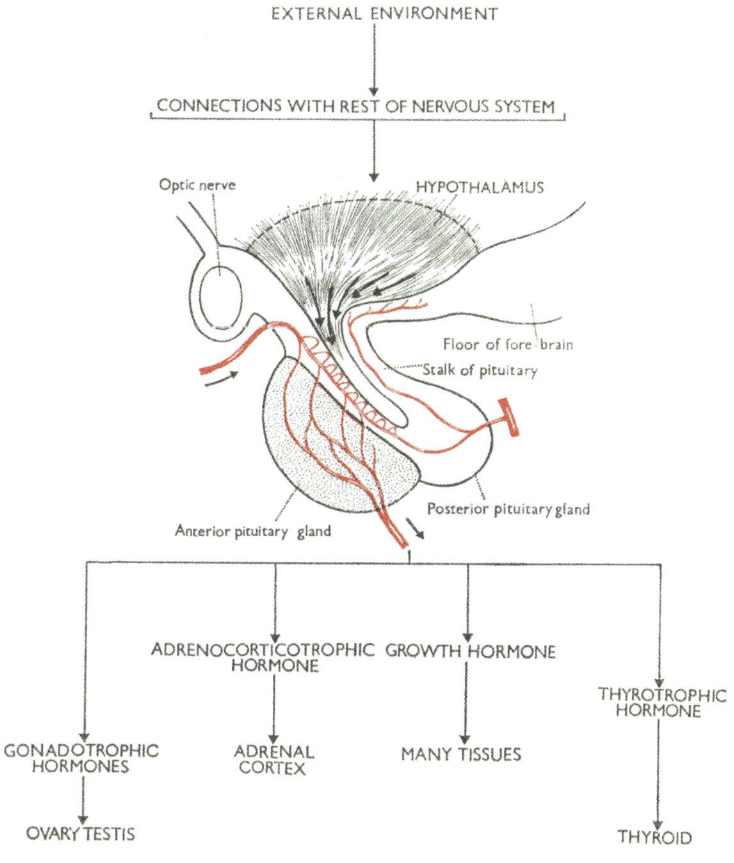

FIG. 4. *The Anterior pituitary gland and its relations with the hypothalamus.* (Vertical section of pituitary and hypothalamus.)

The hypothalamus consists of groups of nerve cells and fibres lying in the floor of the third ventricle of the brain. The approximate limit of this area is indicated by the interrupted line. The hypothalamus has widespread connections with the rest of the brain. A number of nerve fibres converge on the pituitary gland from various regions of the hypothalamus—indicated by arrowed lines. These do not enter the anterior lobe but they terminate on the surface of capillary loops of blood vessels in the stalk of the pituitary. Here they liberate chemical substances which pass into the blood stream and they are transported to the anterior lobe where they influence the activity of the gland. The

[*continued at foot of p. 12*

functions of temperature regulation, appetite, thirst, etc.—with the "master" endocrine gland, the pituitary, which itself regulates the activity of other major endocrine glands (Fig. 4).

Some endocrine glands are themselves sensitive to fluxes in their immediate environment and respond by changes in the rate of hormone production. Thus the parathyroid glands, which exert a major influence in regulating the amount of calcium ions in body fluids, are directly sensitive to changes in [Ca^{++}] of blood and respond to these changes by varying the output of the hormone parathormone. The islets of Langerhans of the pancreas are sensitive to changes in the glucose content of blood and respond by a change in the rate of insulin secretion (page 31). However, for successful adaptation of the organism endocrine glands also need to be able to respond to changes in the external environment of the body. Thus the thyroid gland which has a potent influence on the rate of metabolism of all body cells, is able to regulate metabolism and heat production according to the needs of the organism. Animals exposed to cold show increased thyroid activity with a rise in basal metabolic rate and heat production. Change in superficial and deep body temperature is sensed by the nervous system. In the hypothalamus is a temperature sensing structure of exquisite sensitivity and a fall in deep body temperature causes the hypothalamus to activate secretory activity of the anterior pituitary gland. The pituitary gland responds by an increased output of various hormones including thyroid stimulating hormone and adrenocorticotrophic hormone. The thyroid gland releases increased amounts of hormones and these trigger a rise in metabolic activity of all body cells with its associated increase in the rate of heat production. The increased activity of the adrenal cortex following the release of adrenocorticotrophic hormone results in an increased ability of the organism to withstand the stresses of exposure to cold. If the pituitary gland is separated

---

connection between the hypothalamus is thus a special vascular link—the hypothalimico—hypophysial postal system—which transports activators from the hypothalamus to the anterior pituitary.

Note that this system has two capillary beds interposed between the artery and vein, one in the pituitary stalk and one in the gland itself.

Because of this link between nervous system and pituitary gland, the activity of various other glands (ovary, testis, thyroid and adrenal cortex) can be related to conditions in the external environment.

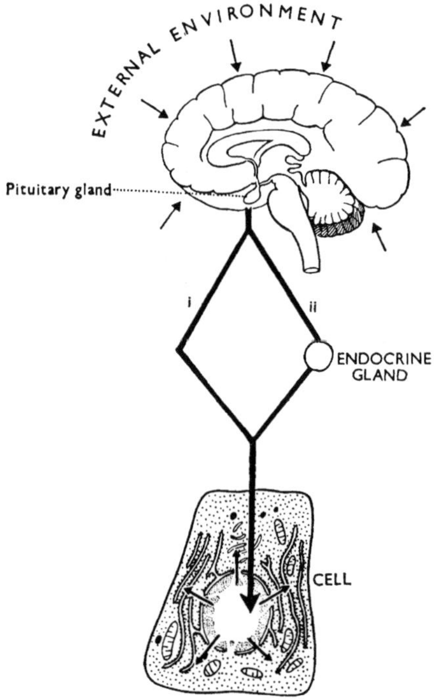

FIG. 5. *Link between the external environment of the organism and the gene.*

The figure illustrates the stages in the activation of cells in response to fluxes in the external environment of the body. These fluxes are initially appreciated by the nervous system which activates the anterior pituitary gland by way of its connections with the hypothalamus. The secretions of the anterior lobe may activate cells directly, e.g. growth hormone ((i) in figure) or indirectly by promoting the secretion of another hormone by an endocrine gland, e.g. cortisone ((ii) in figure). In either case the final chemical messenger alters the activity of the genetic material of the cell. This is manifested by a change in the amount of messenger substances leaving the cell, and thus a change in the metabolic activity of the cell.

---

from the influence of the hypothalamus, either by cutting the pituitary stalk or by transplanting the pituitary to another region of the body, then the nervous system can no longer influence the gland. This results in a failure of adaptation of glands such as the thyroid, adrenal, ovary and testis to fluxes in the external environment.

A Link between Environment and Gene. In the past, biologists have tended to clearly separate the two classes of regulators of cell activities, i.e., the internal genetic factors on the one hand and the external neuro-endocrine influences on the other. However, now that it is realized that some hormones influence the cell by altering gene activity then we can visualize a bridge between the external environment of the body and the genetic apparatus, and this bridge is the neuro-endocrine system. This concept is shown diagrammatically in Figure 5.

CHAPTER 3

Hormones and the Cell Membrane: I Acetylcholine

Since the 19th century endocrinologists have studied the way in which endocrine mechanisms regulate the activities of the cells of various tissues so as to produce a highly integrated organism which is adapted to the external environment. Hormones have been found to be concerned in the regulation of very many body functions including reproduction, the metabolism of carbohydrate, protein, fat, sodium, calcium, phosphorus and water, in growth and maturation, temperature regulation and in adaptation of the organism to stresses such as muscular exercise, injury and infection. The situation is complex. There are many hormones in the body and each hormone may produce a variety of effects. We cannot here study the role of all hormones in all body processes. We shall instead consider a few hormones in detail, examining how the contribution of modern cellular studies and molecular biology can provide us with at least partial answers to the fundamental problem of how particular hormones produce their effects in the organism.

The Cell: Possible Sites of Action of Hormones

Figure 6 shows a schematized cell indicating some cell components which have been discovered by the various forms of microscopy—the light microscope, phase-contrast microscope, electron microscope. Provided that a hormone molecule has withstood the vicissitudes of transport from the endocrine gland to the target cells (and these vicissitudes are many and include inactivation by enzymes in blood and tissues, binding by plasma proteins, excretion by the liver and kidney) there are obviously many possible sites inside the cell with which the hormone might react. It is by no means certain that all hormones do in fact enter the target cells, and some may exert only a peripheral action on the cell. The number of potential sites at which the hormone molecules can act is of course much greater than is obvious by looking at Figure 6 because the ultimate interaction of hormone and cell is at a molecular level. The possible nature and sites of such molecular interactions will be the subject of subsequent discussions.

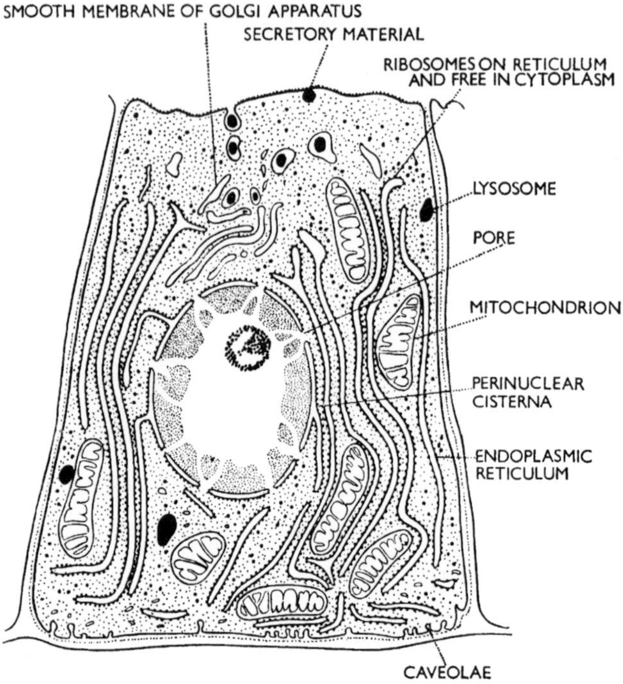

FIG. 6. *Contents of a "typical cell".*

Note the following:

Cell membrane. On the lower part of the figure the membrane shows pits or caveolae which are thought to be involved in the uptake of substances into the cell. On the upper part of the figure there are secretory granules being extruded from the cell.

Endoplasmic reticulum. The synthesis of protein for secretion by the cell is associated with arrangement of the *ribosomes* on flattened membranes sacs which communicate with one another to form a network. The secretion is believed to pass through the channels of the reticulum into the smooth membranes of the *Golgi apparatus.* This is thought to act by collecting secretion, surrounding it by a membrane, and the resulting membrane bound vesicle containing secretion leaves the Golgi apparatus and moves to the cell membrane. Here the membranes fuse and the secretion passes out of the cell.

Mitochondria contain the enzymes involved in the step-wise oxidation of energy producing substrates such as fatty acids and glucose.

Ribosomes occur both free in the cytoplasm and on the surface of the

[*continued at foot of p. 17*

A single hormone may produce diverse effects in different tissues. Indeed even in a single cell a particular hormone may produce a variety of effects. If we consider a single effect of a hormone this may be the result of action of the hormone at one of several loci in the cell. These loci are listed below:—

1. the surface membrane of the cell
2. enzymes (intracellular or located in the cell membrane)
3. genetic material inside the nucleus
4. cellular organelles, e.g., mitochondria, ribosomes.

The Action of Hormones on the Surface Membrane of Cells

The cell membrane (plasma membrane) has been regarded as a bimolecular sheet of lipo-protein, the sheets of lipo-protein being separate by a finite gap (Fig. 7). The molecules of the phospholipid compound have a long hydrophobic chain with a small polar, hydrophilic group at one end. In the cell membrane the non-polar groups (hydrophobic) are opposed, and the polar group turned outwards. In some cells the lipid layer is probably more complex than this and consists of compound layers of cholesterol and phospholipid.

This structure readily explains certain properties of the membrane, e.g., that the permeability of the membrane to a particular substance is in general related to the solubility of the substance in lipids. Certain substances, however, penetrate the cell membrane very rapidly and it is assumed that the free entry of small particles less than 7A° is due to the presence of pores in the cell membrane having a diameter of about 7·1A°.

The structure of the cell membrane as described above obviously provides little scope for the action of hormones and indeed this description of the cell membrane must be regarded as a gross simplification of the real nature of the membrane. The membrane

---

endoplasmic material. Contains RNA and protein. Sites of protein synthesis.

Nucleus surrounded by a flattened sac of endoplasmic reticulum, the outer surface of which carries ribosomes. Pores penetrate the nuclear membrane. The nucleus contains the chromatin material with its DNA which instructs the activities of the cell. The chromosomes are not obvious in this non-dividing cell.

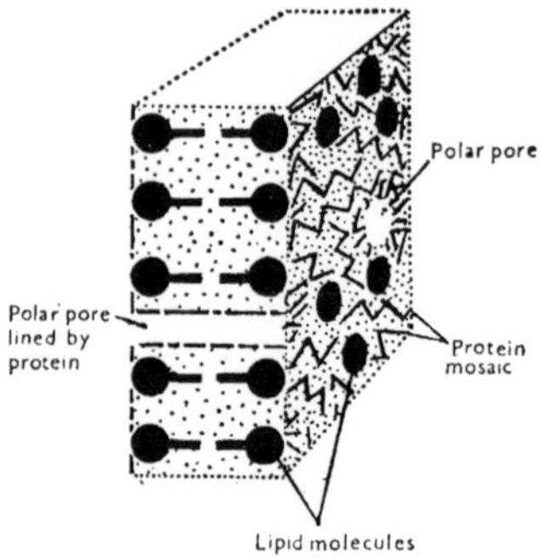

FIG. 7. *Diagram illustrating the structure of the plasma membrane.* The membrane consists of two layers of protein with lipid molecules dispersed between the protein molecules. Two polar pores are shown in the diagram.

---

is a dynamic not a static structure; many cells show invaginations of the membrane (caveolae) which are thought to be concerned with the transport of some substances into the interior of the cell. It has been suggested that some molecules become bound to specific sites on the outer surface of the membrane and that this infolds to form a vesicle which breaks away from the surface membrane to the interior of the cell, where it may release its contents or unite with intracellular membranes (Fig. 6).

The cell membrane also contains other mechanisms which transport materials between the extracellular fluid and the intracellular contents, e.g., glucose and the sodium ion. In addition the cell membrane contains active sites which react with substances in the extracellular fluid. These active sites are called *receptors*, and although we may not yet know the nature of the receptor site we know that they must exist, and indeed in some instances their distribution can be visualized.

Acetylcholine and the Cell Membrane

Distribution, and Function of Acetylcholine. In 1921 Otto Loewi provided the first clear evidence that nerves exert their efforts upon tissues by releasing a special chemical substance at the nerve terminal. Loewi inserted a fine tube into the heart of a frog through which he ran a salt solution. The saline solution issuing from this heart was allowed to come into contact with a second frog heart. He then stimulated the vagus nerve supplying the first heart. As was expected, stimulation of the vagus caused a slowing of the heart beat and ultimately stopped it. However, in addition he noted that there was a more gradully developing arrest of the beating of the second heart. This he attributed to the release of some inhibitory chemical substance from the first heart on stimulation of the vagus nerve. This substance was carried over to the second heart in the saline solution. Figure 8 shows the results of

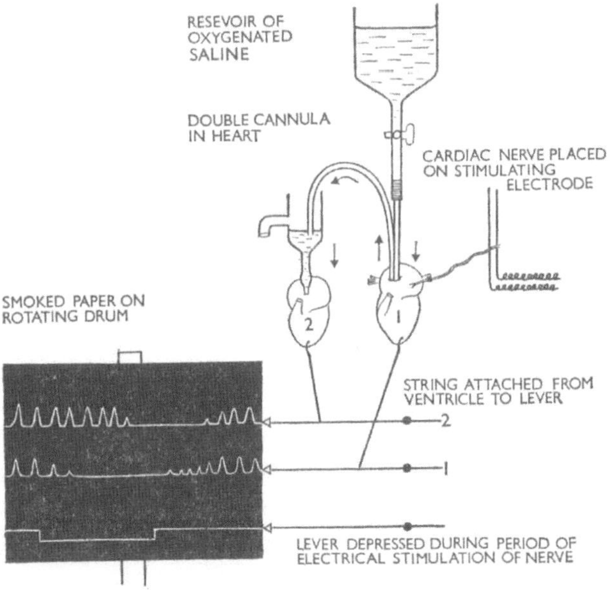

FIG. 8. *Apparatus showing a modification of Loewi's technique illustrating the release of vagusstoff.*

Isolated heart 1 is perfused with saline solution which leaves heart 1 to perfuse heart 2. Stimulation of the cardiac nerve to heart 1 results in rapid cessation of beating of the heart (lower kymograph trace) followed after an interval by the cessation of beating of heart 2. The duration of electrical stimulation of the nerve is indicated by depression of the lowest line on the kymograph.

---

an experiment repeating Loewi's observations. Loewi called this substance "Vagusstoff" (vagus substance). This substance has now been identified as acetylcholine.

Many other investigations have confirmed that a chemical substance is involved in the transmission of the nervous impulse to many body tissues. Acetylcholine is widely distributed within the central and peripheral parts of the nervous system and is involved in the transmission of the nerve impulse at many different sites. Acetylcholine is responsible for transmission of the nerve impulse at the neuromuscular junction in the somatic nervous system, at terminals of all parasympathetic nerve fibres (e.g. salivary glands, heart, bowel, bladder) and at some sympathetic nerve terminals (e.g. sweat glands), at autonomic ganglia (both parasympathetic and sympathetic), and may also be involved in interneuronal transmission within the central nervous system. At the terminals of sympathetic nerve fibres another substance (noradrenaline) is usually involved in transmission.

Storage and Release of Acetylcholine. The acetylcholine which is present in tissues is present in a physiologically inactive non-diffusible form. The atrium of the heart, for example, contains about 500 times the amount of acetylcholine needed to cause a maximal effect on the atrium if it were all released at once. Most of the acetylcholine in the body is present in nerve fibres. Prolonged stimulation of the nerve fibres causes a decrease in the acetylcholine content of the tissue. Electron microscope studies of nerve terminals have demonstrated the presence of large numbers of vesicles—synaptic vesicles—measuring only a few hundred Angstrom units across. It is these vesicles which contain the bound, physiologically inactive acetylcholine. The arrival of a nerve impulse at the nerve terminal causes the release of a minute quantity of acetylcholine (only 10^{-17}♍.) from the synaptic vesicles into the cytoplasm of the nerve fibre from which it passes out of the nerve fibre. Physiological evidence indicates that acetylcholine is released from a large number (200) of sites in the form of small packets of the hormone containing a constant number of molecules. This evidence agrees with the known facts about the distribution of acetylcholine in the synaptic vesicles. Resynthesis of the neurotransmitter substance acetylcholine is a continuous process and the stores do not decline unless there is prolonged stimulation of the nerves.

Inactivation of Acetylcholine Released from Nerve Terminals. The action of acetylcholine at the nerve terminal is a brief event—at the neuromuscular junction in striated muscle a matter of milliseconds. This is because free acetylcholine is inactivated by the enzyme acetylcholinesterase which hydrolyses the molecule into choline and acetate, both of which are physiologically inactive. At some nerve terminals this enzyme is present in high concentrations.

Acetylcholine and Skeletal Neuromuscular Transmission

The action of acetylcholine in transmission of the nerve impulse from somatic motor nerves to striated muscle has been the subject of intensive and fruitful investigation.

The Motor End-plate. The relationship between a motor nerve fibre and a skeletal muscle fibre is shown in Figure 9. As the nerve

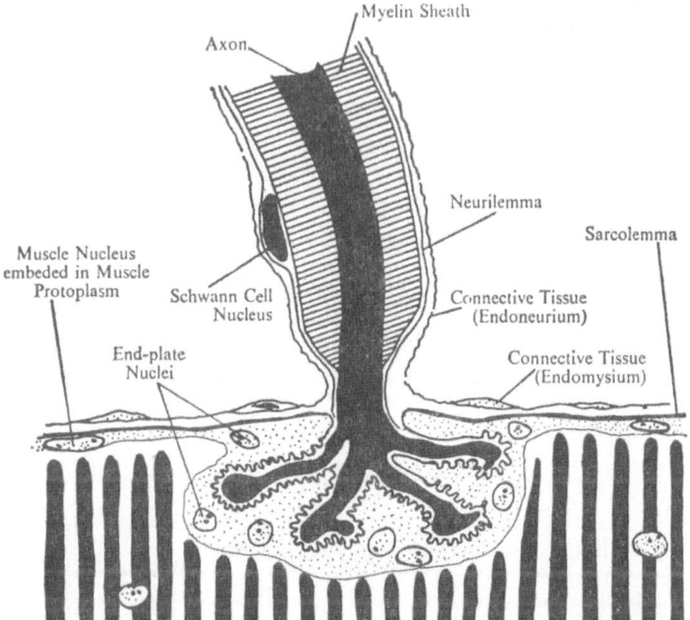

FIG. 9. *Diagram of a motor end-plate.*

approaches the muscle the myelin sheath is lost and naked nerve terminals become embedded in an area of muscle cytoplasm. This relationship is called the motor end-plate. Note that there is a minute but finite gap between the nerve terminals and the muscle cytoplasm. This gap between two excitable cells is called a synapse. The post synaptic membrane, i.e., the muscle membrane underlying the nerve terminals, is a complex highly infolded region which thereby presents a large surface area.

Transmission at the Motor End-plate. Conduction of the nerve impulse along the motor nerve is a propagated electrical event at the level of the membrane of the fibre. There are two possible ways in which this event in the nerve fibre could induce the changes in the muscle which lead to contraction. One possibility is that the nerve fibre could supply enough electrical current to the muscle fibre to reduce the transmembrane potential to the threshold level (page 26). The diameter of the muscle fibre is so much larger than that of the nerve terminal that it is highly unlikely that the nerve could supply enough current to depolarize a membrane at least 100x larger than itself.

The second possibility is that the arrival of the nerve impulse causes the release of acetylcholine which is responsible for the electrical changes in the muscle membrane which lead to contraction of the fibres. Certainly the electrical stimulation of a motor nerve to skeletal muscle results in the appearance of detectable amounts of free acetylcholine, and the injection of a minute quantity of acetylcholine into the artery of a prepared muscle results in a brief twitch of the muscle which resembles the response to electrical stimulation of the motor nerve.

The Events at the Muscle Membrane. The chemical composition of the interior of cells differs markedly from that of the interstitial fluid bathing the cell. The intracellular fluid is characterized by a much higher concentration of potassium ions and a lower concentration of sodium ions than the interstitial fluid. The negative anions which accompany the positive sodium ions of the interstitial fluid are mainly chloride ions. Inside the cell the anions consist mainly of proteins in their anionic form. Separating these two fluids, the intracellular and interstitial fluid, is the cell membrane. The diffusion of substances across this membrane is much slower than their diffusion in water. However, there is a net tendency for potassium ions to diffuse out of the cell and for sodium ions to diffuse into the cell, the ions moving along their concentration gradients. The membrane is much more permeable [x100] to potassium than sodium ions. Positively charged potassium ions thus diffuse out of the cell, through the cell membrane. They are not accompanied by negatively charged anions from inside the cell because these are too large to pass through the cell membrane. As potassium ions diffuse out of the cell the interior of the cell becomes increasingly electrochemically negative to the outside, because of the loss of positively charged particles. This electrochemical gradient, created by the diffusion of potassium ions out of the cell, actually tends to attract potassium ions back into the cell. Thus the potassium ions which diffuse out of the cell along their concentration gradient are held close to the cell membrane by electrochemical attraction. An equilibrium is set up between the tendency for potassium ions to leave and re-enter the cell. At this equilibrium there is a potential difference across the cell membrane which amounts, in striated muscle cells, to 90 millivolts, the interior of the cell being electrochemically negative.

Although the cell membrane is much less permeable to sodium than to potassium ions there is a net tendency for the intracellular sodium to come into equilibrium with the extracellular sodium. Sodium ions enter the cell along both a concentration gradient and along an electrochemical gradient. The intracellular sodium concentration is, however, kept at low levels because of some active mechanism in the cell membrane which extrudes sodium ions from the interior of the cell into the extracellular fluid.

The above features are characteristic of all cells. In muscle and nerve cells there is the additional feature of excitability in which changes in the external environment of the cell of a chemical or electrical nature can lead to alterations in the permeability of the membrane to ions, which results in changes in the membrane potential.

In muscles and nerves this change in the properties of the membrane is rapidly propagated to adjacent regions of the cell membrane—the phenomenon of conductivity. This propagated change in membrane properties is called the impulse and is manifested electrically as the action potential. At rest the nerve or muscle cell maintains a fairly steady electrical potential across the cell membrane. When the excitable cell is stimulated, either by chemicals or by passing an electric current outwards through the membrane then there is a marked change in the membrane potential. The membrane potential rapidly falls to zero and for a short time the potential is reversed so that the inside becomes positive to the outside. This change is propagated along the membrane so that successive regions of the cell become depolar-

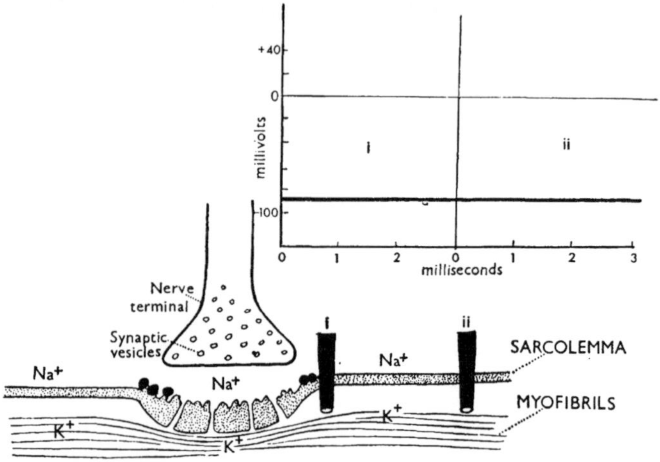

FIG. 10. *Schematic representation of events at the motor end-plate.* In all three figures are shown the nerve terminal with its synaptic vesicles, the folded synaptic membrane containing pores and bearing the enzyme acetylcholinesterase (●). Two micro electrodes are inserted into the muscle fibre. One is located near the motor endplate and another some distance away from the end-plate. Each electrode is also connected to an indifferent electrode (not shown) in the extracellular fluid. A potential difference is recorded between the indifferent and probing electrodes (i and ii). Above the electrodes is shown a record of the potential difference between the indifferent and probing electrodes recorded over a period of a few milliseconds (abscissa). The ordinate shows the recorded potential difference between the indifferent and probing electrodes expressed in millivolts.

1. Fig. 1 shows the state of affairs at rest. No acetylcholine is present in the synaptic space. Na^+ and K^+ refer to the dominant extra- and intra-cellular ions respectively. The recorded potential difference between indifferent and probing electrodes is steady at about 90 mV, the interior of the cell being electrochemically negative to the extracellular fluid.

2. Fig. 2 shows the effect of arrival of a nerve impulse at the motor end-plate. Acetylcholine is released from synaptic vesicles into the synaptic space where it reacts with the postsynaptic membrane causing an increase in pore diameter. The permeability to all ions increases and the transmembrane potential falls (i). When this reaches a critical magnitude the local current flow in the adjacent membrane increases suddenly and an action potential is generated (ii) when the membrane potential falls rapidly to zero and then the polarity is reversed as sodium ions accumulate in the muscle cell.

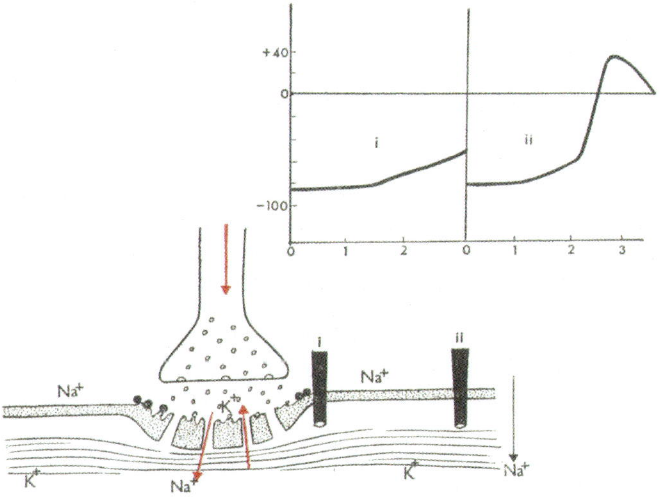

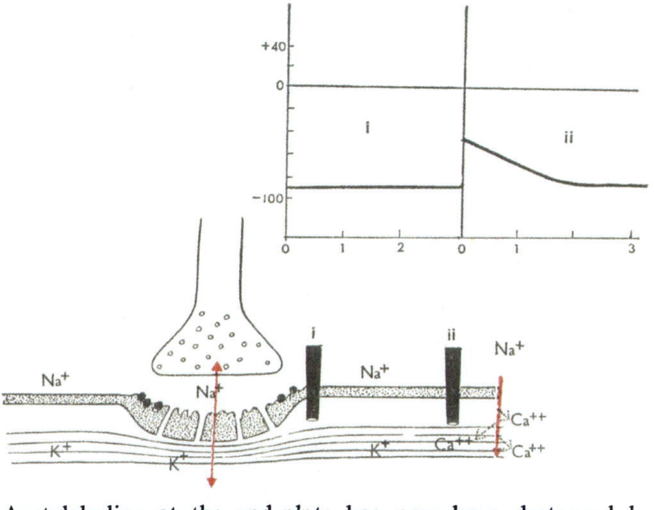

3. Acetylcholine at the end-plate has now been destroyed by the cholinesterase. Excess sodium ions are extruded from the cell and the transmembrane potential returns to normal levels. This process also occurs along the length of the cell membrane behind the front of the action potential. The electrical record at (ii) shows a return to normal transmembrane potential. The depolarization is carried inwards along the endoplasmic reticulum and is associated with the liberation of free calcium ions in the interior of the cell.

ized. Behind the front of the nerve impulse the normal transmembrane potential is restored.

The fall in the membrane potential to zero in the generation of the action potential is due to a sudden increase in the permeability of the membrane to sodium ions (50x increase). Sodium ions move across the cell membrane into the cell interior. They move in along both an electrochemical and a concentration gradient. This influx of positively charged sodium ions into the cell abolishes the electronegativity of the cell interior which for a brief period becomes positive, because of a net excess of positively charged ions. The increase in the permeability of the membrane to sodium ions is a very brief event, lasting only a millisecond or so, and the recovery is associated with a change from a sodium-permeable to a potassium-permeable membrane, the normal state of the resting cell. In the recovery process excess sodium ions are extracted from the cell by the sodium pump mechanism.

The Action of Acetylcholine on the Motor End-plate. The action of acetylcholine on the muscle membrane at the end-plate (the postsynaptic membrane) is to cause an increase in the permeability of the membrane to all ions which are free in the intracellular and extracellular fluids. Such an effect could be due to an increase in size of pores through the membrane so that all ions could cross the membrane more readily. These ions move along their concentration gradients, sodium ions entering the cell and potassium ions leaving the cell. This action should "short-circuit" the membrane to zero potential since the resting electrical potential across the membrane is due to a *gradient* of charged ions. However the action of acetylcholine is so brief that the membrane is never "short-circuited" to zero potential. The depolarization of the end-plate is only partial. The recorded electrical change is called the end-plate potential (e.p.p.). The e.p.p. varies in size depending on the amount of acetylcholine released which in turn depends on the frequency of nerve stimulation.

The fall in the electrical potential at the endplate sets up local circuit flow in adjacent regions of the membrane. Depolarization thus spreads passively into adjacent regions of the muscle membrane. When this reaches a critical magnitude there is a sudden brief increase in the permeability of the muscle membrane to sodium ions. Areas adjacent to the endplate become completely depolarized and a wave of depolarization—the action potential—sweeps along the muscle membrane. This depolarization may spread inwards into the interior of the muscle cell along specialized areas of the endoplasmic reticulum. The link between the electrical events of the muscle membrane and the contractile mechanism itself appears to be a release of calcium ions from bound sites on the endoplasmic reticulum. The inward flux of free calcium into the cytoplasm activates an enzyme concerned in the initial phase of contraction—myosin ATPase. The sequence of events in neuromuscular transmission are summarized in Figure 10.

Acetylcholine Receptors on the Muscle Membrane

We know that on the skeletal muscle membrane, underlying the motor endplate, there are receptors which react with the chemical transmitter acetylcholine which is released from the nerve terminals. The interaction of the acetylcholine with receptors triggers off permeability changes in the muscle membrane resulting in electrical changes which eventually activate the contractile mechanism. Not all areas of the muscle membrane contains these receptors. This is clear from experiments in which acetylcholine is applied to different parts of the muscle membrane and the electrical and mechanical results are observed.

Demonstration of Receptor Sites. The receptor sites on the muscle membrane can be visualized using an ingenious technique. This technique is based on the fact that in addition to acetylcholine certain other substances, because of their chemical similarity, can attach themselves to acetylcholine receptors. The Indian arrow poison, curare, does so. This substance is sufficiently similar to acetylcholine to be attracted to and unite with the receptors. It is not, however, sufficiently similar to trigger off the events which normally follow occupation of the receptors by acetylcholine. The curare molecules are attached to the receptors for some time and, unlike acetylcholine, they are not destroyed by the enzyme choline-esterase which is present in abundance at the motor endplate to ensure that the action of acetylcholine is brief. Thus curare produces paralysis of voluntary muscle because it occupies the acetylcholine receptors which are now unable to accept the neurotransmitter.

Curare can be "labelled" with radioactive carbon. If this radioactive curare is applied to a muscle then the distribution of receptors can be mapped by studying the distribution of spots of radioactivity, and this can be conveniently done by the technique of autoradiography. Figure 11 shows the results of this technique after mouse diaphragm—a very thin sheet of muscle—had been exposed to radioactive C_{14} calabash curare. After exposure to the drug the muscle was spread out on to a photographic film and left until the emitted radiation had affected the film (a matter of several months). When the film was developed, darkened areas showed the effect of radiation emitted from the curare, thus localizing the receptors for acetylcholine. Compare the distribution of blackening with the distribution of motor end-plates shown by another technique.

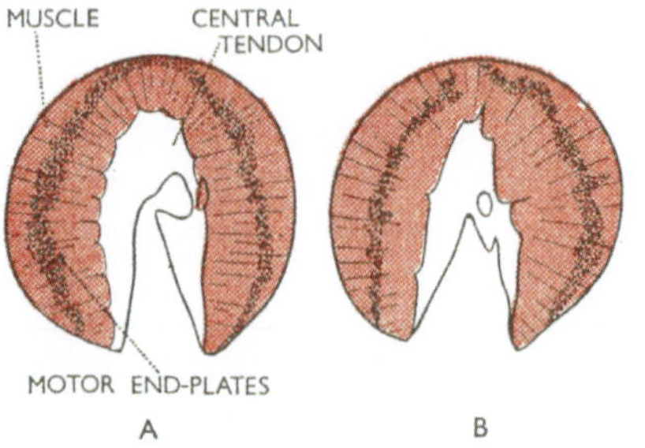

FIG. 11. *Distribution of motor end-plates in mouse diaphragm (A) and sites of radioactivity after exposure to* 14*C—curarine. Figures drawn from photographs by Waser (1962).*

The motor end-plates of mouse diaphragm lie in a circular band around the central tendon. In Figure A the end-plates were stained by a technique which stains the acetylcholinesterase which is concentrated at the endplate. B shows the distribution of radioactivity on the diaphragm after exposure to radioactive curare. It is obvious that this has been taken up by the motor end-plates, and at higher magnification than shown in the figure individual end-plates could be distinguished, having a diameter of only 16–35 microns.

It is possible in these experiments to make an estimation of the number of curare molecules which are bound to each end-plate. The degree of blackening of the autoradiograph film can be measured by a densitometer and compared to the blackening of discs of gelatine of similar thickness to diaphragm and containing different amounts of radioactive curare. These studies gave an estimate of $2{\cdot}6 \times 10^6$ molecules of radiocurare per motor end-plate which agrees closely with estimates of the number of molecules of acetylcholine (6 × 10^6) which are needed for synaptic transmission. This figure of 2·6 × 10^6 is quite small when compared with the large area of the infolded postsynaptic membrane of the motor end-plate. Probably less than 1% of the postsynaptic membrane is occupied by receptors. Presuming a regular structure of the postsynaptic membrane then the receptors are few and widely distributed.

The Effect of Receptor Occupation. This kind of experiment demonstrates in a dramatic fashion that chemical activators do in fact become attached to specific areas of the cell membrane in producing their effects. Of course, we are still left with the problem of how the union of acetylcholine with its receptor triggers off those changes in membrane permeability to ions, which results in a change in transmembrane electrical potential ultimately leading to muscular contraction.

Several simple models have been produced which incorporate known facts about neuromuscular transmission. One such model by Waser is shown in Figure 12.

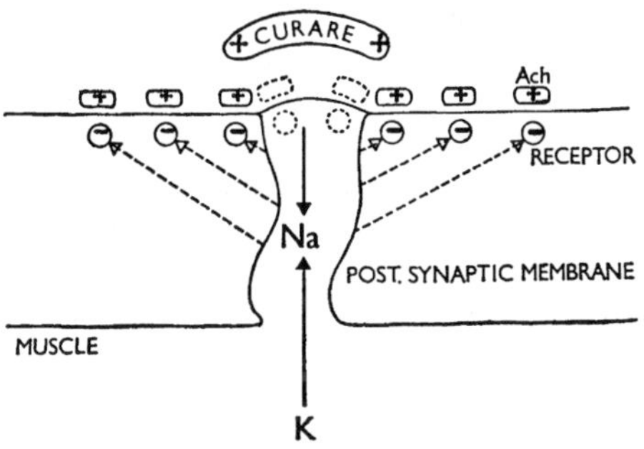

FIG 12. *A hypothetical scheme for the action of acetylcholine.* (*Modified. from Waser 1962.*)

The molecules of acetylcholine are shown on the surface of the postsynaptic membrane with their positive charge attracted to the negative charge on the receptors. It is suggested that the union of acetylcholine with the receptor results in forces (dotted lines) which dilate the channel through the postsynaptic membrane, increasing the permeability of the membrane to ions (Na, K) which move along their concentration gradients.

The action of curare is also illustrated. This positively charged molecule unites with acetylcholine receptors and occludes the pore.

Isolation of Receptor Substance. There have been various attempts to isolate the receptor substance for acetylcholine. For this purpose the electric organ of the eel has been used, which contains large amounts of acetylcholine. In 1959 two workers independently reported that they had isolated a protein from the electric tissue which showed many of the properties of the receptor substance in the intact muscle in life. However, it will be extremely difficult to prove that this protein is the receptor in the tissue. The extraction of the protein from the tissue inevitably must break attaching forces which bind the receptor to the postsynaptic membrane. It is possible that the active protein which has been extracted is in fact part of the receptor substance but with its functional groups intact.

Indirect Effects of Activators on the Cell Membrane. It must be emphasized that the fact that the application of a hormone to a tissue produces changes in the functional characterstics of the cell membrane does not necessarily imply a *direct* effect on the membrane. Thus, although the hormone aldosterone promotes the transport of water across the isolated amphibian bladder autoradiographic studies show that the hormone becomes concentrated in the nuclei of bladder cells, not in the cell membrane. Further, substances which are capable of blocking the responses of the genetic material of the cell to hormones also block the effect of aldosterone on water transport across the cell membrane. It is clear that although aldosterone produces changes in the activity of the cell membrane it does so by acting at a deeper level in the cell, in fact at the level of the gene.

CHAPTER 4

Hormones and the Cell Membrane: II Insulin

The hormone insulin is synthesized and stored in certain cells of the Islets of Langerhans of the pancreas called beta cells (Fig. 13). The hormone is a polypeptide. The most potent stimulus which

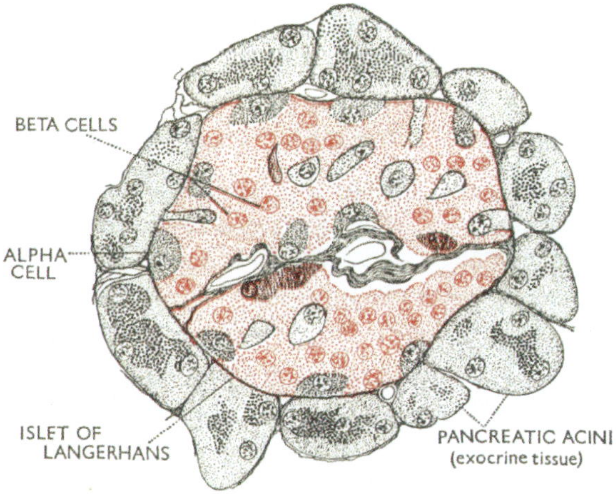

FIG. 13. *Drawing of part of pancreas showing an islet of Langerhans embedded in the exocrine tissue.*

causes a release of insulin from the beta cells of the Islets of Langerhans is a rise in the concentration of glucose in blood, such as occurs after a meal when the products of digestion are entering the blood stream.

The Multiple Actions of Insulin

Insulin produces a variety of effects in the animal. The classical effect of the hormone is that it causes a fall in the concentration of glucose in blood. If sufficient insulin is injected into an animal then the concentration of glucose in the blood falls to such a low level that the central nervous system is deprived of its major source of energy (i.e. glucose) and abnormal brain activity occurs which results in generalized convulsions which may be fatal. This effect of insulin is the result of the action of the hormone at different sites and on different processes in the organism.

In the liver, insulin has two main effects. First, it increases the rate at which glucose is converted to the storage polysaccharide glycogen. Second, it slows down the process whereby the liver manufactures glucose from non carbohydrate precursors such as the non-nitrogenous part of amino acids and glycerol (a process called gluconeogenêsis). Both of these effects on the liver result in a fall in the amounts of circulating glucose. In addition to these effects upon the liver, insulin also acts on other tissues. It promotes an increased rate of uptake of glucose by the mass of fat and muscle tissues of the body. In fat cells this increased glucose uptake is followed by an increase in the rate of synthesis of fatty acids and neutral fats from glucose metabolism. In muscle cells there is an increase in the rate of glycogen synthesis. All of these effects of insulin contribute to the fall in the concentration of circulating glucose, but it is by no means clear that these various actions of insulin are due to similar mechanisms or occur at similar loci in liver, muscle and fat cells.

In addition to these various effects on the metabolism of carbohydrate, insulin is also intimately involved in protein metabolism and thus on growth. Exposure of isolated tissue, e.g., muscle to the hormone results in an increase in the rate of incorporation of amino acids into protein molecules. This effect on protein synthesis is not necessarily related to the effect of insulin on carbohydrate metabolism and in isolated tissues it can be observed even in the absence of glucose in the incubating medium

According to our present state of understanding these various effects of insulin cannot be explained by a single mechanism of action. The action of insulin on protein metabolism is probably due to an effect of the hormone at the level of the gene. This view is based on the fact that insulin produces marked changes in the amount of RNA in cells, an effect which can be blocked by use of the agent actinomycin (page 63). In contrast the effect of insulin on the uptake of glucose by muscle and fat cells—for reasons which will be discussed later—is regarded as an effect at the cell membrane.

The Action of Insulin at the Cell Membrane

The Access of Glucose to Cells. Before looking at the evidence for the site of action of insulin on carbohydrate metabolism in fat and muscle cells, let us first consider some of the factors which are involved in the uptake and metabolism of glucose by cells. The transfer of glucose from the circulating blood into the interior of the cell involves movement of glucose molecules across two membranes, first the capillary wall which gives access to the interstitial fluid of the tissues and, second the cell membrane which gives access to the interior of the cell.

The transfer of glucose from the blood to the fluid which bathes the cells, takes place under the influence of simple physical forces. The capillary wall presents no barrier to the diffusion of glucose molecules and the concentration of glucose in the interstitial fluid rapidly comes into equilibrium with the glucose of blood when this is artificially altered.

However the movement of glucose from the interstitial fluid into the cells seems to depend upon a special mechanism located in the cell membrane. This is a system of *active transport* of sugar molecules across the membrane. It is suggested that the membrane contains a substance called a carrier which is capable of entering into a reversible combination with glucose. It is supposed that the carrier is capable of oscillating or diffusing across the cell membrane and that association with the carrier at the outer surface of the cell membrane and disassociation from it at the internal surface of the membrane is the only way in which glucose can cross the cell membrane. This mechanism then provides a means by which glucose is transferred along the concentration gradient (Fig. 14).

Not all body cells which transport glucose across the cell membrane are affected by insulin (e.g. red blood cell). However, in many body cells, insulin increases the rate at which glucose is transported into the cell. It has been suggested that the insulin molecule might enter into reversible combination with the carrier molecule altering the properties of the carrier molecule so as to increase the efficiency of transport. One way in which this might occur is by a reduction in the affinity of the carrier molecule for glucose—i.e. a reduction in the tendency for the carrier and glucose to combine. Reduction of the affinity between glucose and the carrier molecule would allow a more ready release of glucose from the carrier at the internal surface of the cell membrane. It would also, of course, reduce the amount of glucose picked up at the outer cell membrane. It may seem paradoxical that a reduction in the affinity of the carrier for glucose could increase the rate of

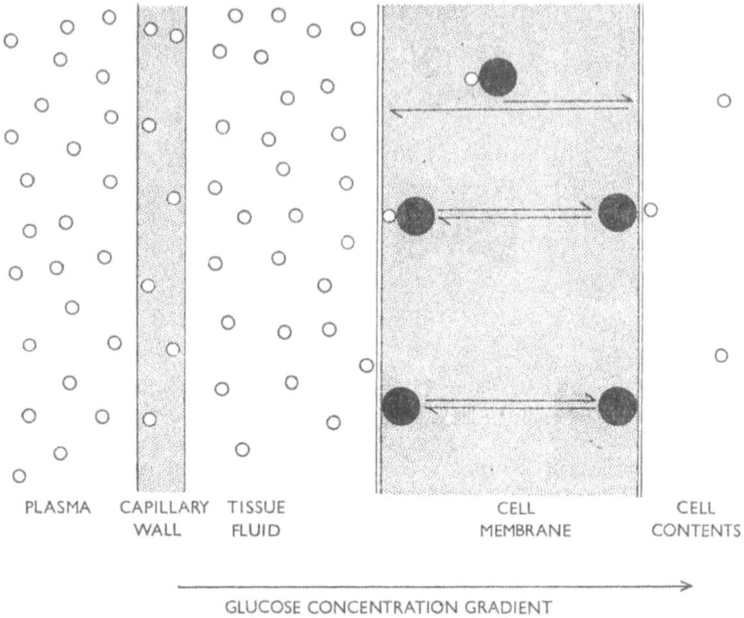

FIG. 14. *Diagram showing the membranes separating glucose in plasma from the cell contents.*

The capillary wall presents no barrier to exchange of glucose between plasma and tissue fluid and the concentration of glucose (○) is the same in the two spaces. The cell membrane is a barrier to the free movement of glucose from tissue fluid into the cells and glucose molecules can only gain access by combination with a carrier molecule (●) which oscillates across the cell membrane. This scheme particularly applies to muscle and fat cells. Some cells, e.g. liver and brain, do no show such a membrane barrier to the movement of glucose.

---

glucose transport. If a carrier had infinite affinity for glucose then little or no glucose would be released into the cell, however low the concentration of glucose in the cell. Conversely a carrier with a low affinity for glucose would also result in minimal transport since little glucose would be accepted on the outer surface of the membrane. Obviously there is an optimal affinity of the carrier for the glucose molecule for efficient transport of glucose. Certainly little glucose is transported in the absence of insulin and the hormone may act by modifying the properties of the carrier molecule to bring its affinity for glucose into an optimal range.

Recent studies of this transport system show that its behaviour is predictable from the law of mass action (i.e. it shows Michaelis—Menton Kinetics). The transport mechanism also shows stereo-specificity—that is only sugars with particular steric configurations are transported.

Experimental Evidence for the Site of Action of Insulin. The amount of glucose in muscle cells and the effect of insulin.

As early as 1939 a worker called Lundsgaard obtained experimental results which led him to the conclusion that insulin influences the mechanisms which transfer glucose from thc extracellular fluid into the interior of the cell. These studies were made with animals from which the abdominal viscera including the pancreas had been removed. These animals were thus unable to secrete insulin, unable to store or generate glucose because they had no liver, and since they had no kidneys could not excrete administered glucose. He injected glucose into these animals by way of a vein and then, after various intervals, he removed samples of blood and muscle for analysis of their glucose content. Muscle tissue consists of various components, principally the muscle cells themselves and the surrounding interstitial fluid. An analysis of a piece of muscle for glucose content gives no information whatsoever of the distribution of the glucose in the original tissue. The glucose could have been in the muscle cells, in the interstitial fluid, or in both. One can, however, come to a fairly reliable estimate of the amount of glucose in the interstitial fluid of muscle. The glucose of interstitial fluid is in equilibrium with that of blood plasma so that if one knows the concentration of glucose in plasma at the time the muscle tissue was removed for analysis then this provides information about the *concentration* of glucose in the interstitial fluid of muscle tissue. However, the important measurement we need is the *total amount* of glucose in the interstitial fluid of the muscle sample and this can only be determined by having an additional piece of information—the volume of interstitial fluid in muscle. This can be measured in various indirect ways to give a fairly reliable estimate. Knowing the volume of interstitial fluid in muscle tissue (measured indirectly on the same sample of muscle used for glucose analysis) and its glucose concentration (which is equal to the plasma glucose concentration at the time of sampling), then one can assess the total amount of glucose present in the interstitial fluid of a piece of muscle tissue. If the whole muscle sample contains a greater total amount of glucose than is calculated to be present in the interstitial fluid of the piece of muscle then one can readily assume that the extra glucose must be present inside the muscle cells.

Lundsgaard found, however, that the amount of glucose present in muscle samples did not exceed that which he calculated to be present in the tissue fluid of the muscle. He came to the conclusion that the amount of free glucose in muscle cells must be very low. He formed the opinion that the transfer of glucose from the extracellular fluid into cells must be an important limiting factor in the utilization of glucose by cells.

When he injected insulin in addition to the glucose into the animals he found a rise in the amount of glucose in muscle tissues which he concluded was due to an uptake of glucose by the muscle cells.

Lundsgaard's views on the action of insulin were confirmed by Levine (1949). Levine studied the effect of insulin on the distribution of the sugar galactose in dogs from which the abdominal viscera had been removed. Galactose is a sugar which is not utilized by the body and this removes some of the uncertainties of experiments using glucose. Levine studied the "volume of distribution" of galactose following injection of the sugar into the blood stream. Using the relationship between the known amount of substance added to an unknown volume of fluid and the final concentration of substance in the fluid one can calculate the volume of the fluid.

$$V = \frac{A}{C}$$

V = Volume
A = Amount of substance added
C = Final concentration of substance

After the injection of galactose into the circulating blood there is a progressive fall in the concentration of galactose in the blood, which is due to the movement of galactose into other fluid compartments of the body—principally the interstitial fluid (tissue fluid) and the intracellular fluid. Levine found that after injection of galactose into the blood stream the final volume of distribution of galactose (calculated from the known amount injected and the final blood concentration of galactose after sufficient time had elapsed for galactose to come into equilibrium in the various fluid compartments) was equal to that of the volume of extracellular fluid (i.e. plasma volume and tissue fluid volume). When insulin was injected with the galactose there was a fall in the concentration of galactose in the blood and thus the calculated volume of distribution increased. In the absence of metabolism of galactose by the tissues this could only be accounted for by a movement of galactose into the cells (Fig. 15).

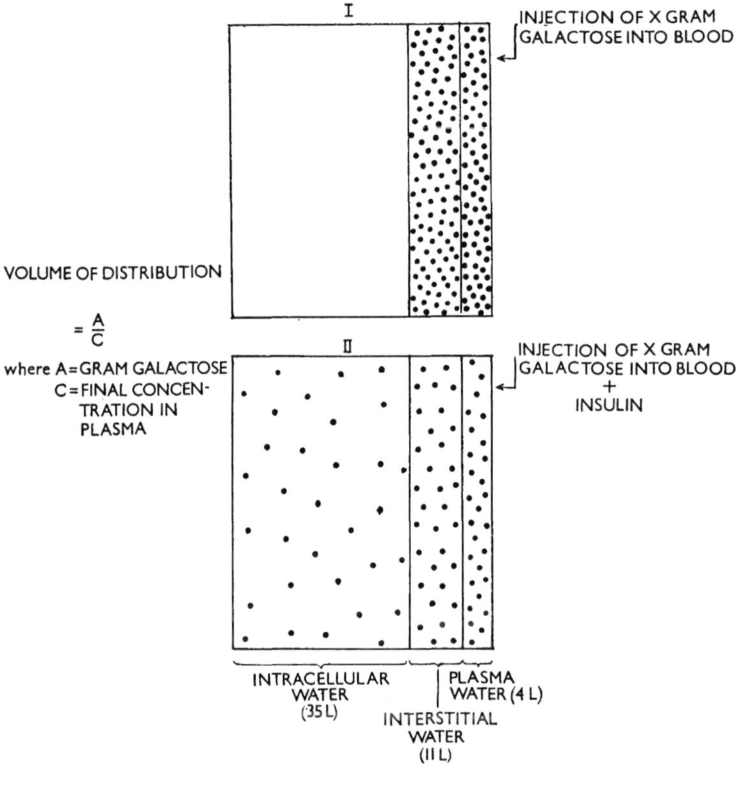

FIG. 15. *Diagram showing the effect of insulin on the distribution of a sugar (galactose) in the various fluid compartments of the body.*

Note the different sizes of the fluid compartments. The number of litres for each compartment refers to average figures for man.

In I, galactose is injected into a vein and becomes, in the absence of insulin, distributed predominantly in extracellular fluids (plasma and interstitial water). The volume of distribution is given by the formula.

[*continued at foot of p. 38.*

It must be mentioned that another concept of the mode of action of insulin on carbohydrate metabolism was accepted for some time. A worker named Cori found evidence to support the view that insulin promoted the action of the intracellular enzyme hexokinase. Hexokinase is involved in the initial phosphorylation of the molecule of glucose before it enters one of the various metabolic paths in the cell. If insulin activated this enzyme then insulin should decrease the concentration of free intracellular glucose, since the sugar should be more rapidly incorporated into the metabolism of the cell. In fact insulin promotes an *increase* in the concentration of intracellular glucose. This hypothesis has now been generally disregarded. The studies of Lundsgaard and Levine have been confirmed by many workers using a variety of isolated organs, tissues and sugars. It can readily be shown that after treatment with insulin there is an increase in the intracellular concentration of various sugars when these are present in the solution perfusing an organ. We can conclude that insulin stimulates the efficiency of the cell membrane transport system for glucose. There is little information of the exact nature of this effect on the membrane (page 33). Exposure of a piece of tissue to insulin results in binding of the hormone to the membrane. Thus the action of the hormone persists in tissues which have been briefly exposed to hormone and then washed free of the hormone. Radioactive insulin is similarly bound to cells and the study of the various fractions of the cell after homogenizing the tissue show that the radioactivity is mainly located in the cell membrane fraction. It has been suggested that insulin binding by tissues involves an interchanging reaction between the thiol groups (–SH) on the cell surface and the disulphide bond (S–S) in the insulin molecule.

The action of insulin on glucose transport is not inhibited by substances which reduce cell metabolism or which prevent the response of genetic material to activators.

Significance of the Action of Insulin on Glucose Transport. Before leaving the question of the action of insulin we must briefly

---

In II, the same amount of galactose is injected but with insulin. The final concentration of galactose in plasma is less than in I, i.e. the volume of distribution is greater. This is because insulin promotes the transport of galactose into the cells. Note that the concentration of galactose in cell water does not equal that in extracellular fluids.

consider the biological significance of this effect. The concentration of glucose in blood in healthy young men rarely rises above 100mg% even after eating a meal rich in carbohydrate and the concentration falls below 60mg% only after prolonged fasting or very severe exercise. In view of the many possible pathways of glucose utilization (e.g. formation of glycogen, fat, some amino acids, oxidative metabolism) this level of control is truly remarkable. The achievement of this control is due to the co-operation of many systems of glucose production and utilization, many of which are capable of being modulated by hormones in accordance with the requirements of the body (see also Chapter 6). Blood glucose must be maintained above a certain minimal level to ensure normal function of the central nervous system which is virtually entirely dependent upon glucose as a source of energy. If supplies of glucose are poor, e.g. in fasting, then the decline in insulin production which automatically follows results in a diversion of glucose from metabolism by fat and muscle cells. In the absence of insulin very little glucose reaches these tissues—but they have alternative sources of energy in the form of oxidation of fatty acids. Glucose is thus spared for the requirements of the central nervous system.

In the process of urine formation large volumes of a protein free fluid is filtered into the tubular tissues of the kidney. This filtrate of plasma contains the same glucose content as plasma. During the elaboration of the urine by the renal tubules this glucose is reabsorbed into the bloodstream. There is, however, a finite limit to glucose reabsorption by the renal tubules and this limit is normally reached when the glucose content of plasma rises above 180mg%. At this point glucose begins to appear in the urine and increases in amount with rising plasma content. This indeed occurs in the patient suffering from diabetes mellitus who does not secrete sufficient insulin. Normally a rise in plasma glucose content is followed by an increase in insulin production which results in a diversion of glucose into fat and muscle cells. This glucose is laid down as glycogen in muscle or utilized in the synthesis of fatty acids and neutral fat in fat cells, and is not lost in the urine.

CHAPTER 5

The Action of Hormones on Intracellular Enzymes: Adrenaline

To distinguish between a hormone acting on the cell membrane rather than on an intracellular enzyme is, to a degree, artificial and rests upon older concepts of cell structure in which the cell was regarded as a blob of cytoplasm surrounded by a limiting cell membrane of uniform character. We now know that membranes are not limited to the surface of the cell but ramify through the entire cell structure as components of the endoplasmic reticulum, Golgi apparatus and mitochondria. Moreover, a large number of enzymes are associated with membranes. Thus the enzymes associated with electron transport and oxidative phosphorylation are associated in a membrane structure (the mitochondrian). The enzymes concerned with the formation of the nucleotide 3′,5′-AMP are associated with the outer cell membrane. This latter system will be discussed in some detail in the present chapter.

Adrenaline

The Chromaffine Cell. The adrenal gland consists of two components, an outer cortex and an inner medulla; these are quite distinct in their structure, function and their origin in the embryo. The medulla forms the greyish core of the gland, containing groups or cords of cells surrounded by blood vessels. The medulla produces adrenaline, together with a closely related hormone, noradrenaline (Fig. 16). The secretory cells are called chromaffine cells because of the darkening of the tissues after exposure to solutions of chromic acid or bichromate.

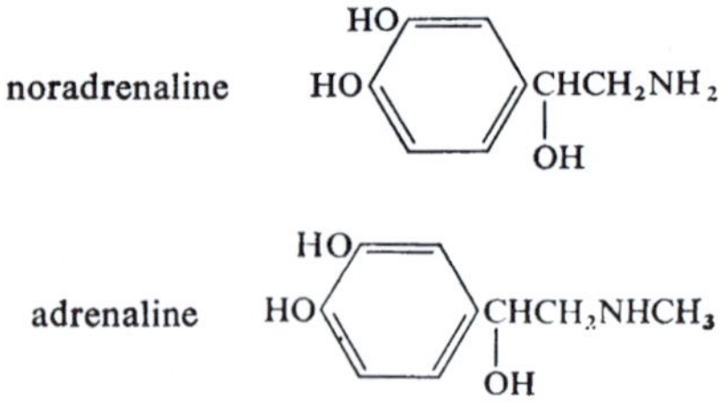

FIG. 16. *The chemical structure of adrenaline and noradrenaline.*

Chromaffine cells are not, however, restricted to the adrenal medulla. They are widely distributed in the body and occur, for example, on the dorsal abdominal wall and in the pelvic nervous plexuses. This widespread distribution of chromaffine cells is related to their mode of origin. They arise in development as primitive nerve cells mainly in the thoracic region of the spinal cord. From here they migrate along the course of the sympathetic

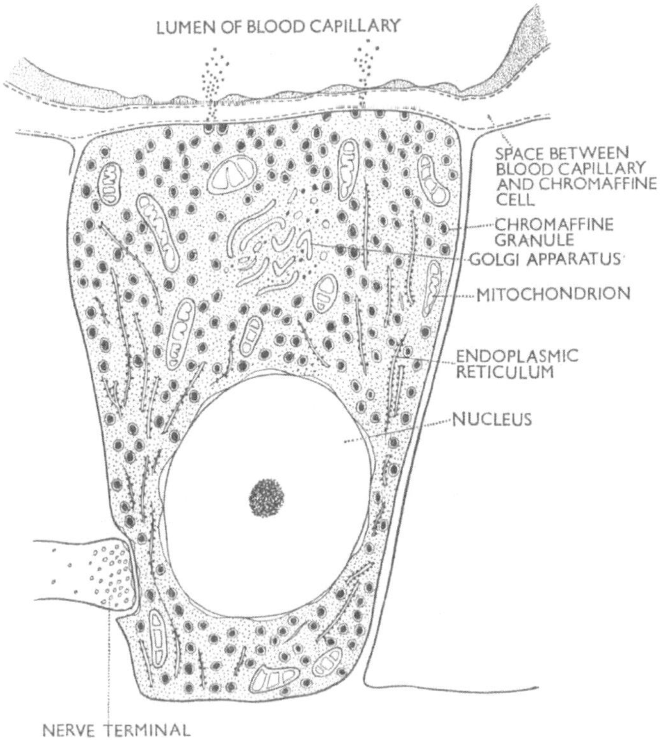

FIG. 17. *Diagram showing typical structure of a chromaffine cell.*
The chromaffine granules containing the hormone appear to arise in the Golgi apparatus. The membrane bound granules are shown approaching the cell membrane where the two membranes appear to fuse releasing the contents which finally reach the lumen of blood capillaries. Note the "patchy" nature of the capillary wall adjacent to the secreting cell, facilitating the diffusion of the hormone into the bloodstream.

nerves. Some of the cells which reach the dorsal abdominal wall become associated with adrenocortical cells to form the core of the adrenal medulla. Other primitive nerve cells never reach the dorsal abdominal wall and yet others migrate further into the pelvis.

Chromaffine cells, both adreno-medullary and extramedullary, are similar in structure. The secretory activity of all chromaffine cells is regulated by the sympathetic nervous system (Fig. 17). In fact chromaffine cells are highly modified post-ganglionic neurones of the sympathetic nervous system which are specialized to produce, in large quantities, the chemical transmitter normally released at the sympathetic nerve terminals. However, the chromaffine cells which are embedded in the adrenal gland are peculiar in that they produce adrenaline rather than the closely related compound noradrenaline. The reason for this difference is not clear but it has been suggested that the blood flowing into the medulla from the adrenal cortex, containing large amounts of adrenocortical hormones, induces the presence of the appropriate methylating enzyme which converts noradrenaline into adrenaline. Some cells of the adrenal medulla do still produce noradrenaline and the blood supply of these cells is derived from the adrenal artery rather than from the venous blood draining inwards from the adrenal cortex.

Physiological Effects of Adrenaline and Noradrenaline

Of the two hormones adrenaline is in general a more potent compound, and it also has some effects which are not produced by noradrenaline. This may be of significance in the evolution of the relationship between adrenocortical and chromaffine tissue, i.e. the relationship may confer some advantages on the organism since it results in the secretion of a preponderance of adrenaline. Comparing the activities of equal molecular doses adrenaline is in general 4–8 $\times$ more potent than noradrenaline.

The actions of the hormones are widespread in the body and are involved in the emergency adaptation of the body to stress, permitting a more efficient performance in "fight or flight". The changes produced by adrenaline and noradrenaline include an increase in heart rate and the vigour of cardiac contraction, which result in an increase in the rate of ejection of blood from the ventricles. This in itself will cause a rise in blood pressure. But an additional factor in the rise in blood pressure which follows administration of the hormones is a widespread contraction of small arteries—the arterioles—in many tissues. Adrenaline differs from noradrenaline in its effect on arterioles. Although adrenaline causes vasoconstriction in many organs, e.g. skin and abdominal viscera, it dilates the blood vessels of the voluntary muscles and brain. This means that blood is shunted from skin and viscera into those areas which vitally need a good blood supply if the animal is to succeed in "fight or flight". Other body systems are also alerted to emergency by the hormones, in particular the nervous system. Noradrenaline is a general vasoconstrictor of all arterioles and does not therefore have this "selective shunting" effect of adrenaline.

Metabolic Effects

In addition to effects on the cardio-vascular system and nervous system there is also a prominent metabolic response. This is shown by an early rise in the concentration of glucose and lactic acid in blood following the administration of adrenaline. Associated with these changes there is a decline in the glycogen content of liver and muscles. Adrenaline is very effective in producing these effects on carbohydrate metabolism and does so in doses which are less than those needed to raise the blood pressure. The significance of these effects is that the body is provided with a ready source of energy—glucose—which may be required in "fight or flight". Noradrenaline, however, has minimal effects on carbohydrate metabolism. We can now examine in some detail how these effects of adrenaline on carbohydrate metabolism are produced.

Glycogen Metabolism of Liver and Muscle

Glycogen is a polysaccharide in which glucose molecules—or rather glycosyl units—are joined together by α1,4 links to form straight chains. These straight chains are further joined together by α-1,6 links to produce a branching structure (Fig. 19). Units of glucose can be added to, or removed from the large branching molecule—which thus has a very variable molecular weight. In man the liver contains about 75 G of glycogen, enough to supply the body with glucose for 12 hours in the absence of other sources of glucose.

The breakdown of the glycogen molecule in the liver which results in the formation of free glucose takes place in three stages. In the first stage glycogen reacts with inorganic phosphate in the presence of the enzyme phosphorylase. This reaction breaks the α1,4 links between glycosyl units of the straight chains and liberates glucose 1-phosphate. This process proceeds along the straight chains of the molecule until an α-1,6 link of a branching point is reached. This link is then broken under the influence of another enzyme—debranching enzyme—which then permits the enzyme phosphorylase to continue along the straight chain. Glucose-1-P produced by the action of phosphorylase can then be converted to glucose-6-phosphate by the enzyme phosphogluco-mutase. Glucose-6-phosphate can then enter various metabolic pathways in the cell—it can enter pathways of oxidative meta-bolism, undergo oxidation to gluconic acid-6-phosphate which is

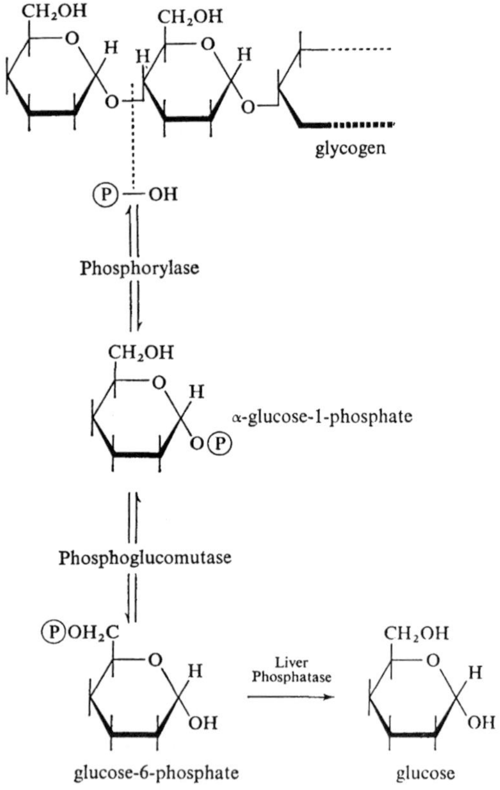

FIG. 18. *Stages in the breakdown of glycogen.*

an important step in the formation of the pentose sugar D-ribose, be incorporated into the glycogen molecule and so on. In addition free glucose can be derived from glucose-6-phosphate in the liver by means of an irreversible reaction catalysed by the enzyme phosphatase. Although many other tissues, e.g. cardiac and skeletal muscle contain glycogen they do not contain this phosphatase. This means that these tissues cannot provide glucose for the general body economy and the glucose-6-phosphate derived from glycogen breakdown is used for the tissues own metabolic requirements. These stages in glycogen hydrolysis are shown in Figure 18.

The rate of glucose production for glycogen in the liver depends upon the amount of enzyme phosphorylase in liver cells. The amount of phosphorylase depends upon the rates of inactivation and reactivation of the enzyme. This enzyme phosphorylase itself is being continually inactivated by means of another specific phorphorylase. Reactivation of the phosphorylase depends upon its rephosphorylation by ATP in the presence of the enzyme phorphoryl kinase. This enzyme is activated by a cyclic nucleotide, adenosine -3′,5′-phosphate (Fig. 19).

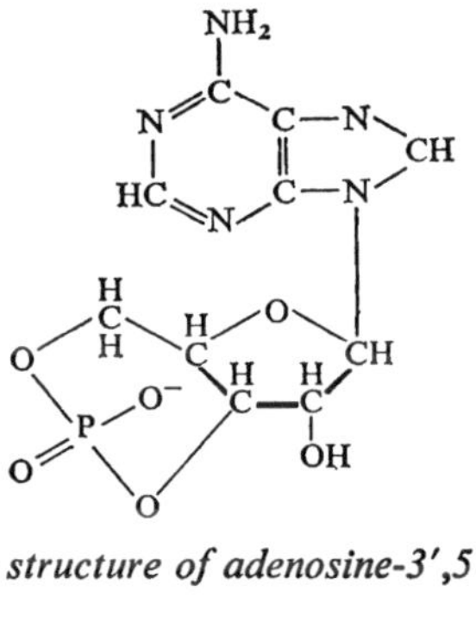

FIG. 19. *The structure of adenosine-3′,5′-phosphate.*

Cyclic adenosine -3′,5′-phosphate (3,5-AMP) like the enzyme phosphorylase, is in a constant state of flux dependent upon rates of production and destruction of the nucleotide. The nucleotide is produced in the presence of various factors including ATP, the enzyme cyclase and magnesium ions. It is at this level that adrenaline acts. Adrenaline stimulates the rate of synthesis of cyclic adenosine -3′,5′-phosphate (3,5-AMP) which activates phosphoryl kinase which in turn activates phosphorylase (Fig. 20).

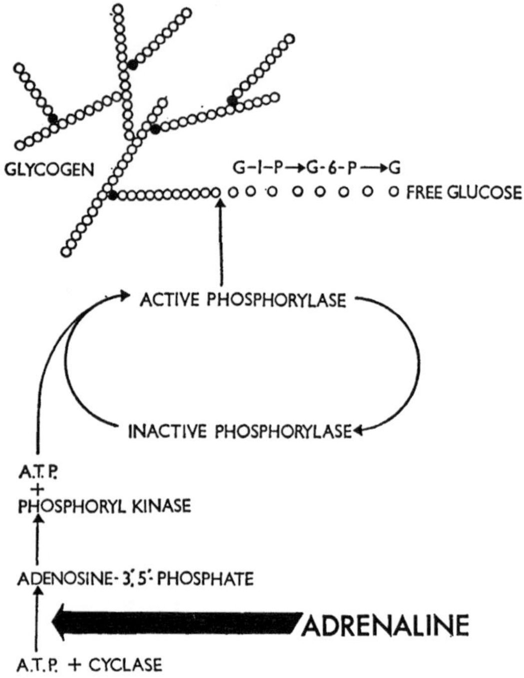

FIG. 20. *The site of action of adrenaline in glycogenolysis.*

After the injection of adrenaline into the whole animal one can detect an increased amount of active phosphorylase in liver cells and a decrease in the glycogen content of liver cells and a rise in glucose concentration in blood.

Second Messenger Concept

In addition to liver cells, other body tissues respond to adrenaline by the production of increased amounts of the nucleotide 3′,5′-AMP. In skeletal muscle cells 3′,5′-AMP has an effect on the enzyme phosphorylase similar to that in liver. However, because of a difference in the enzyme content of liver and muscle cells, in particular a lack of the enzyme glucose-6-phosphatase in muscle, the effect of the hormone in the two tissues is different.

In both tissues glycogen stores become depleted because of the activity of phosphorylase, but in muscle the glucose-6-phosphate is not transformed into glucose. Instead there is an increased rate of metabolism of glucose-6-phosphate which results in lactic acid formation. Other body tissues also respond to adrenaline by increased 3′,5′-AMP production. Such tissues include adipose tissue, uterus, intestinal smooth muscle, spleen, lung and brain.

The action of cyclic 3′,5′-AMP in the body is not restricted to the enzyme phosphorylase. A large number of metabolic processes have been found to be influenced by this nucleotide and include the transport of sugar by epithelial cells of the thyroid gland, water permeability of toad bladder, synthesis of liver proteins and steroid hormones. Various enzyme systems are also activated by the nucleotide. 3′,5′-AMP obviously is of wide significance in the regulation of metabolic processes. Moreover the action of the hormone adrenaline on the synthesis of 3′,3′-AMP is by no means specific. Many other hormones produce similar changes in their target tissues (Table 2).

These observations have given rise to the concept that regulation of the activity of a target organ by an endocrine gland may involve a two (or more) messenger system. The first messenger is the hormone released from a particular endocrine gland following appropriate stimulation of the gland. This first messenger is then transported in the bloodstream to the target organ where it promotes the synthesis of a second messenger which then alters the metabolism of the target tissue to produce the effect (Fig. 21). The only second messenger known at present is the nucleotide cyclic 3′,5′-AMP.

We must try to provide some explanation for the fact that many different hormones produce an identical second messenger but they do so only in their own target tissues. Thus the action of thyroid stimulating hormone on cyclic AMP production is specific to the thyroid gland as is the action of luteinizing hormone on the corpus luteum. In target tissues the "receptor" for first messenger is adenyl cyclase the enzyme concerned in the formation of cyclic AMP from ATP. It has been suggested that the molecular configuration of adenyl cyclase may vary from tissue to tissue, a particular molecular configuration providing a specific receptor for a particular hormone. Reference to Table 2 will reveal that in some cases more than one hormone will stimulate second messenger formation in a particular tissue. Thus both adrenaline and glucagon stimulate cyclic AMP production in the liver resulting in the breakdown of glycogen. We must presume that adenyl cyclase of liver cells can act as a receptor for the two hormones.

In addition to possible differences in the molecular configuration of adenyl cyclase from tissue to tissue, different configurations having affinity for different hormones, there may also be variations in the other enzymes of the tissue which could account for the

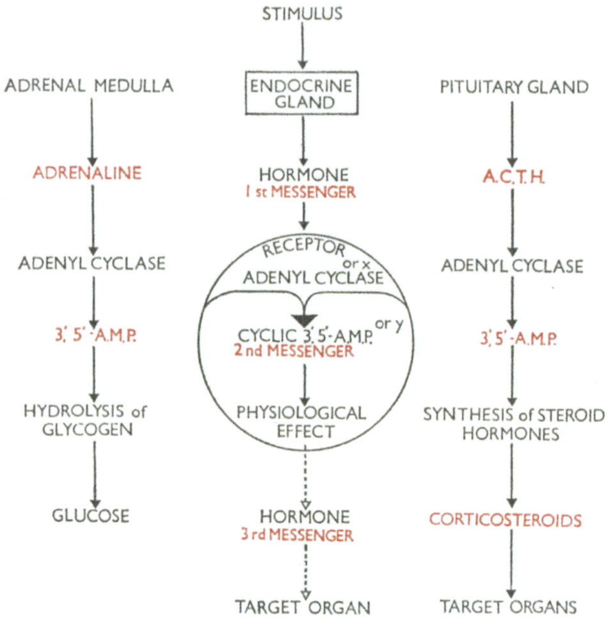

FIG. 21. *Concept of action of hormones by two or three messenger systems.*

In the middle section x + y refer to other possible but unknown substances. On the left is an example of a two messenger system, and on the right a three messenger system.

specificity of action of hormones (first messenger). Thus the hormone adrenaline produces different effects in liver and muscle although there is an identical second messenger in the two tissues.

This difference in response of the two tissues is due to the lack of the enzyme glucose-6-phosphatase in skeletal muscle.

We have said that regulation of the activity of target tissues by an endocrine organ may involve a two messenger system as outlined above. However, a third messenger is known to be present in those tissues which respond to second messenger by the production of another hormone (third messenger) which leaves the organ to act on other tissues. To consider briefly an example of such a three messenger system. The anterior pituitary gland by way of its connections with the central nervous system is activated during conditions of "stress"—that it when an animal is exposed to cold, heat, physical injury and the like, or whenever vital processes such as the regulation of blood pressure or blood glucose content are impaired. The response to such stresses is that the anterior lobe discharges a first messenger—adrenocorticotrophic hormone (ACTH)—into the bloodstream. The target organ for ACTH is the adrenal cortex, which responds by the synthesis of increased amounts of second messenger, the nucleotide 3′,5′-AMP. This second messenger in turn induces changes in the metabolism of adrenocortical cells resulting in an outpouring of various hormones (e.g. cortisol, corticosterone) which act as third messenger. This third messenger is transported in the blood to the various target tissues, the response of which determine an increased ability to withstand stress.

TABLE 2

Hormones which Produce an Accumulation of Cyclic 3′,5′-AMP in Tissues

Hormone	*Tissues*
Adrenaline	Liver, skeletal muscle, adipose tissue, smooth muscle of gut and uterus, lung, brain
Histamine	Gastric mucosa
Acetylcholine	Heart
Glucagon	Liver, adipose tissue
ACTH	Adrenal cortex, adipose tissue
Thyroid stimulating hormone	Thyroid
Luteinizing hormone	Corpus luteum
Vasopressin (ant-diuretic hormone)	Toad bladder, frog skin, kidney

CHAPTER 6

Hormones Acting at the Level of the Gene: I Cortisone

The Adrenal Glands

The importance of the adrenal glands in the life of mammals has been known since the mid nineteenth century. It has however, taken over a century of scientific investigation to provide partial explanations of how the hormones which are produced by the adrenal cortex exert their effects on cells. For very many years physiologists and biochemists were dominated by the notion that hormones must modify cell metabolism either by acting on the cell membrane or on intracellular enzyme systems. Certainly some hormones do appear to act directly on the cell membrane and we have looked in some detail at the way in which insulin and acetylcholine influence the membrane. However, with few exceptions (notably adrenaline, glucagon, ACTH) the study of the action of hormones on enzyme systems has proved to be rather unrewarding. In a review of the literature on the action of adrenal hormones at the cellular level Professor Bush (1962) had to come to the conclusion that "despite much intensive work on the problem, the action of all classes of steroid hormones at the molecular level remains unexplained". He came to the conclusion that this problem was one of the major challenges of physiology and biochemistry.

Advances were made so rapidly that in the same year that Bush came to the above conclusions it was demonstrated that cortisone caused the synthesis of a particular enzyme in liver cells. Further, evidence was provided that this effect of cortisone was due to an action of the hormone at the level of the gene. Before looking in more detail at this action of cortisone we must consider briefly the structure and physiology of the adrenal cortex.

Structure and Activity of the Adrenal Cortex. The adrenal cortex forms the outer yellowish layer of the adrenal gland. It shows three distinct zones when suitably prepared sections are examined under the microscope (Fig. 22). These zones consist of an outer zona glomerulosa, a middle zona fasciculata and an inner zona reticularis. In terms of function however there are two zones which differ in the nature of hormone production, in their functions in the organism and in the mechanisms which regulate their secretory activity.

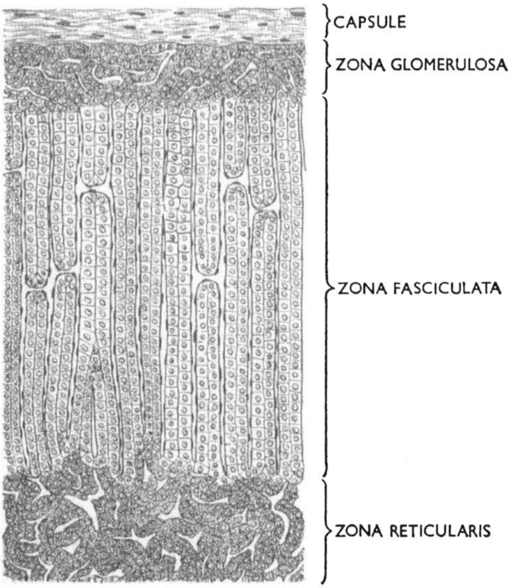

FIG. 22. *The microscopic structure of the adrenal cortex.*

The outer zona glomerulosa is concerned in the production of the hormone aldosterone. This hormone forms part of the body's mechanisms for the regulation of salt and fluid balance. An important site of action of this hormone is the distal convoluted tubules in the kidney where the hormone promotes the reabsorption of sodium ions from the glomerular filtrate in exchange for potassium or hydrogen ions which are secreted by the tubules into the glomerular filtrate. Unlike other zones of the adrenal cortex the glomerulosa remains intact and functional after removal of the anterior lobe of the pituitary gland. The mechanisms which regulate the secretory activity of the glomerulosa are complex and are still the subject of research and debate.

The zona fasciculata and reticularis form a functional unit which secretes mainly the hormones cortisol (hydrocortisone) and corticosterone (cortisone). The cells of the outer zona fasciculata have quantities of lipids in their cytoplasm—the raw material for synthesis of the steriod hormones. The inner densely staining zona reticularis is the actual site of synthesis of the hormone, although under the influence of a strong stimulus the two areas (fasciculata and reticularis) may become unified into a single biosynthetic zone.

The hormones produced by the zona fasciculata and reticularis—principally cortisol and cortisone—are essential for life. If the adrenal cortices are removed the animal may appear normal for some time after it has recovered from the effects of anaesthetic drugs and surgery. Sooner or later, usually a matter of weeks, there is a general deterioration of the condition of the animal. Many disorders appear and these include a loss of appetite, vomiting, diarrhoea, a fall in blood pressure, failure of kidney functions, loss of weight and failure of growth. The amount of glucose in blood may fall to such a low level that convulsions occur and these may be fatal (page 31). Death may occur earlier than anticipated if the animal is exposed to "stress" in any form such as physical injury, exposure to heat or cold, infection or starvation. These various defects in the adrenalectomized animal are not due to the lack of a single effect of adrenocortical hormones. The hormones act at many different sites in the body and are involved in various physiological processes—in the metabolism of carbohydrate, protein and fat, in salt and water regulating mechanisms, in the maintenance of blood pressure, in function of the central nervous system and in the mechanisms of resistance to many forms of "stress". We shall examine in detail only one action of adrenocortical hormones, an action in liver cells which appears to be an effect of the hormone deep in the interior of the cell at the level of the gene.

One clear defect of the animal without adrenal cortices is an inability to maintain normal amounts of glucose in blood, particularly during fasting. In these circumstances the amount of glucose in blood may fall to such a level that disturbances in function of the central nervous system appear, and there may be generalized convulsions due to abnormal electrical activity of the cerebral cortex. In addition to this defect in the regulation of glucose in blood the adrenalectomized animal has poor stores of glycogen in the liver. In the normal fasting animal, supplies of glucose for the needs of the central nervous system are provided by the liver where glucose is manufactured from the non-nitrogenous residues of amino acids and from glycerol—a process called gluconeogenesis.

Contrasting with the effects of removal of the adrenal cortices are the results of administration of large amounts of cortisone or hydrocortisone to normal animals. This treatment causes an increase in the glycogen content of the liver and a rise in the concentration of glucose in the blood. The amount of glucose in blood may be so great that the animal excretes glucose in the urine. This glucose is ultimately derived from gluconeogenesis in liver cells. Because protein is being used as a source of glucose there is an increased rate of breakdown of body protein which in young animals results in retardation of growth.

The Cellular Components on which Cortisone Acts

In order to understand the way in which some hormones such as cortisone can exert their effects by altering the activity of genetic material we must first consider, albeit briefly, what is known of the way in which genes direct the activities of the cell.

The Site of Protein Synthesis in the Cell. The genetic material resides in the nucleus and it controls the activities of the cell by regulating the synthesis of those protein molecules which determine the chemical activity of the cell—that is the enzymes. Enzymes are composed of some or all of the twenty different amino acids, linked together in a unique pattern for each kind of enzyme. These "biological catalysts" of chemical activity are not, however, manufactured in the nucleus of the cell but are synthesized in the cell cytoplasm. It is now clear that the assembly of amino acids to form protein molecules occurs mainly on those minutes particles (15mμ diameter) in the cell which are called ribosomes. These ribosomes occur free in the cytoplasmic matrix of the cell and also attached to the surface of some of the internal membranes of the cell, the endoplasmic reticulum (Fig. 6). The number of ribosomes varies from one cell to another depending upon the stages of growth, cell division and activity in protein synthesis.

The Link between the Nucleus and the Ribosomes. Since the site of protein synthesis is in the cytoplasm of the cell and the genetic material is in the nucleus it seems clear that there must be some link between these two sites, a link which involves the transfer of information from the genes to direct the activities of the ribosomes in protein synthesis.

Evidence for this link—the messenger RNA hypothesis—came originally from experimental studies with bacteria and viruses. One example of such studies is the effect of infection of bacteria with bacteriophages. Bacteriophages (Fig. 23) are very small

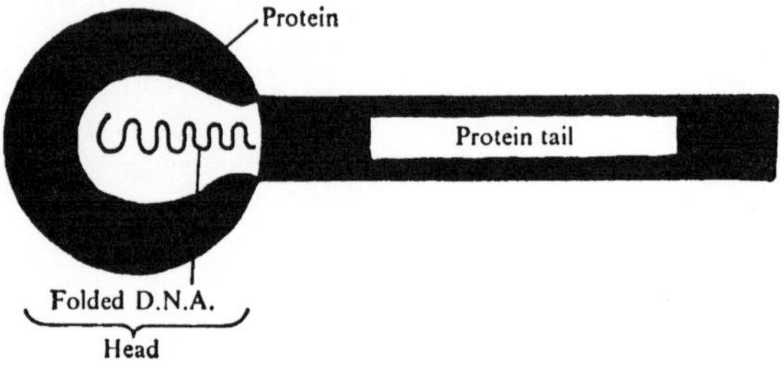

FIG. 23. *Structure of a phage.*

viruses the individual particles of which are shaped like a drumstick. They are peculiar in that their nucleic acid content is entirely composed of desoxyribose nucleic acid (DNA). DNA in other organisms occurs only in the genetic material of the cells, where it forms very long chains in the chromosomes. Bacteriophage contains no ribose nucleic acid (RNA) i.e. nucleic acid in which the sugar unit is ribose not desoxyribose as in DNA. In other organisms RNA is found mainly in the cytoplasm where it occurs in three main forms, messenger RNA, soluble or transfer RNA and ribosomal RNA. The structure of these RNA's may be very different and they perform different functions. *Messenger RNA*, the evidence for which we shall be dealing with shortly, carries the information for protein synthesis from the genes to the ribosomes. This RNA is manufactured on the chromatin where DNA acts as the template for the assembly of the messenger molecule. *Ribosomal RNA*, as its name implies occurs in the ribosomes which are composed of RNA and protein. *Transfer RNA* has a special function in attaching amino acids to messenger RNA during the course of protein synthesis.

To return to experiments with bacteriophages and bacteria. We have seen that the nucleic acid of the bacteriophage consists entirely of DNA, at least in bacteriophage which infect bacterial cells. During the process of infection of the bacterial cell the "phage" sticks to the bacterium by its "tail" and the DNA of the "phage" is injected into the bacterium leaving the "phage" as an empty protein sac. After infection there is a dramatic change in the chemical activities in the bacterium. Synthesis of bacterial protein now ceases and instead the bacterium starts to manufacture proteins of the bacteriophage. It has been found that very shortly after infection a new kind of RNA appears in the bacterium. This new RNA has a structure which indicates that it has been synthesized using "phage" DNA as a template instead of bacterial DNA. To this kind of RNA which is synthesized on the DNA template, and which instructs the ribosomes in the pattern of protein synthesis, is given the name messenger RNA.

In order to understand the significance of this argument we must briefly consider the structure of nucleic acids. Nucleic acids consist of units called nucleotides which are polymerized together to form a chain of units. There are usually only four kinds of nucleotide in a given nucleic acid but nucleic acids are very long chain molecules consisting of hundreds of nucleotides, each of the four nucleotides being repeated many times. Let the nucleotides be called A,B,C,D. Then the structure of a possible nucleic acid might be

ABCDABCDABCDABCDABCD. . . . ABCD

and another might be

AADACBAADACBAADACB. . . . AADACBAADACB

Thus using this four letter code we could produce many different strings of letters. The letters used, their relationship to other letters and the number of times they occur are important in the kind of "word" they produce. So it is with nucleotides and nucleic acids. Although there are only four kinds of nucleotide in a nucleic acid there are many different kinds of nucleic acid.

Each nucleotide consists of a nitrogen containing base (a purine or pyrimidine), a sugar (ribose or desoxyribose) and phosphoric acid. DNA, found mainly in the nucleus of cells, is a stable substance which contains the pyrimidine bases cytosine and thymine and the purine bases adenine and guanine. Ribose nucleic acid is principally cytoplasmic where it is concerned with protein synthesis. The amounts present vary depending upon the extent of protein synthesis. It is far more unstable than nuclear DNA and is being constantly replenished. It contains the pyrimidine bases cytosine and uracil and the purine bases adenine and guanine (Fig. 24). In addition to differences in distribution, chemical composition and stability, the two DNA's differ in their molecular organization.

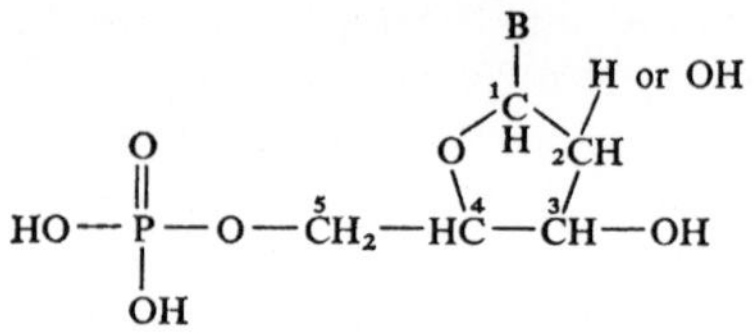

A

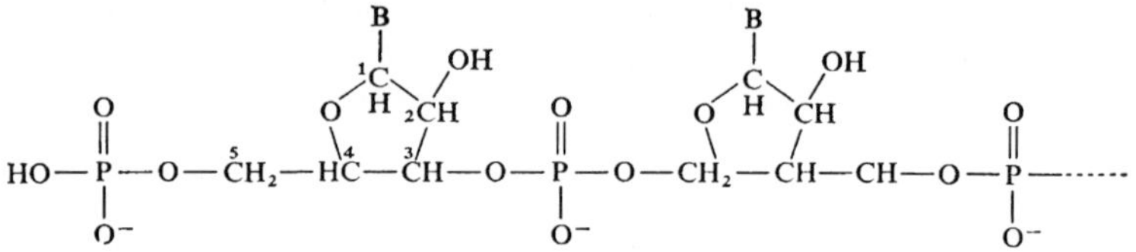

B

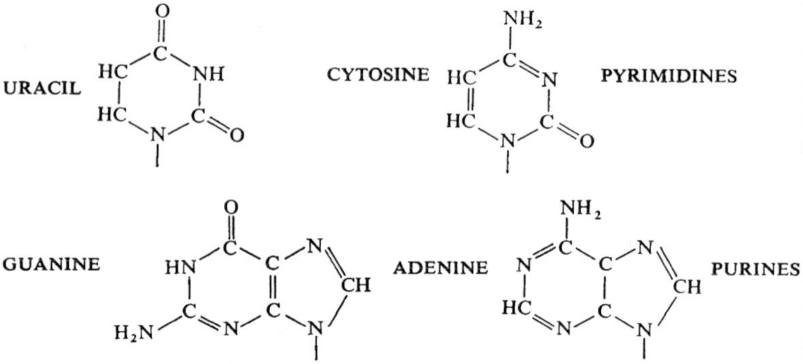

C

FIG. 24. *Structure of nucleic acids.*

A. Shows a unit of nucleic acid structure—a nucleotide, consisting of phosphoric acid, a sugar—desoxribose in DNA (—H at 2) or ribose in RNA (—OH at 2), and a base (B).

B. Part of the backbone of a nucleic acid chain showing union of 2 nucleotides.

C. Formulae of nucleotide bases, the pyrimidines uracil and cytosine and the purines, guanine and adenine.
All four are found in all types of nucleic acid although in RNA the pyrimidine base is uracil while in DNA it is 5-methyl-uracil or thymine (formula not shown).

The molecule of DNA consists of two distinct strands one with one part wound round the other in the form of a double helix (Fig. 25).

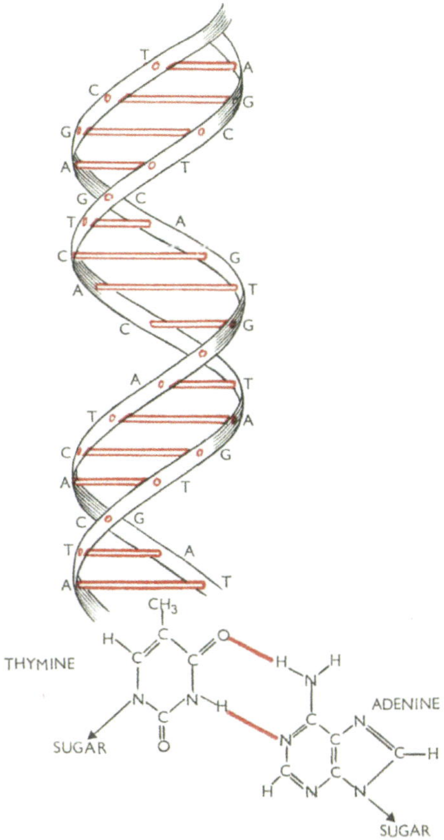

FIG. 25. *Diagram of Watson–Crick model of DNA.*
Showing the double helical structure. The backbone of each member of the helix consists of alternating phosphate and sugar units to which are attached the various bases—thymine (T), adenine (A), guanine (G) and cytosine (C). Complementary bases on the two strands are shown linked by hydrogen bridges. The hydrogen bridges between thymine and adenine are shown in more detail below.

Each strand of the molecule consists of a backbone of alternating desoxyribose sugar and phosphate units. The bases adenine, guanine, cytosine and thymine are attached to this backbone, one base being attached to each sugar unit. The bases of one chain fit neatly into the bases of the partner chain. In order to fit, the bases of one chain must be opposite particular bases of the partner chain. Adenine must be opposite thymine, and guanine must be opposite cytosine. The bases of the two strands are linked together by hydrogen bridges. The nucleic acid chains which coil round each other in the double helix are therefore identical but head in opposite directions. DNA is self replicating and the basic idea is that one strand of the helix acts as a mould and one as a cast. The two strands separate, the mould makes another cast and the cast making another mould. Thus one helix becomes two and the amount of DNA has been doubled. This only occurs during cell division when each daughter cell receives an identical complement of nuclear DNA.

The structure of the molecule of RNA has been much more difficult to elucidate than was the structure of DNA. This is because of the difficulty of preparing RNA in a crystalline form. Molecular structure is worked out by analysing the structure of crystals. In 1962 transfer RNA from yeast cells was isolated in a crystalline form. These are among the shortest nucleic acid molecules (they have about 70 bases strung out along the molecule where DNA has thousands) and the crystalline specimen was not one pure transfer RNA but a mixture of about twenty. The molecules of RNA do not consist of two complementary chains in the form of a double helix as in DNA. Although there is evidence for a double helical structure of some kinds of RNA it appears likely that the nucleotide chain is bent back upon itself with base pairing over part of the molecule.

To return now to our discussion of the evidence for the existence of a special link between nucleus and ribosome in the form of messenger RNA. The evidence from the effect of "phage" infection of bacteria indicated that the switch in the pattern of protein synthesis is associated with the appearance of a new kind of RNA molecule in the bacterium. A study of the sequences of bases on this RNA indicated that it was manufactured on the "phage" DNA not on bacterial DNA.

There is also evidence that in normal bacteria there is a fraction of RNA which has a rapid turnover and which becomes attached to the ribosomes to instruct protein synthesis. This RNA has a base ratio which is more like that of DNA than that of the RNA of the ribosomes. It appears that most of the RNA of the ribosome is "structural" and shows a similar composition in all body cells. To this structural RNA becomes attached a short lived messenger RNA with a base sequence determined by a given length of DNA. The kind of messenger on the ribosomes varies from cell to cell, and it even varies in a single cell thereby determining the fluxes in protein synthesis. In bacteria the messenger RNA has a short existence before it becomes replaced by a similar or different kind of RNA. This state of affairs is not necessarily true of all cells and some experiments have shown that protein synthesis may occur, albeit at a reduced rate, in the absence of chromatin and thus presumably in the absence of new messenger RNA.

We can summarize thus by saying that information is supplied to the ribosomes for the manufacture of a particular enzyme in the form of a molecule of ribose nucleic acid—messenger RNA. This message is written in a four letter code, that is in terms of the sequence of arrangement of the four RNA monomers (uracil, cytosine, guanine and adenine) along the molecule. Messenger RNA is manufactured on the nuclear chromatin using DNA of the chromatin as a template. Thus messenger RNA has a base sequence complementary to the base sequence of a segment of DNA of chromatin. Ribosomes decode this message and translate it in terms of synthesis of protein. We shall not concern ourselves here with the details of what is known of this decoding operation nor of how the peptide chain is assembled on the ribosome. This will be considered in Chapter 9.

The Basis of Cellular Differentiation

Every cell of the body contains the same kind and amount of DNA. However, in different types of cells different types of proteins, different enzymes are produced. In a particular differentiated cell the enzymes which are produced thus form only a small fraction of the number of enzymes which the cell is potentially capable of producing. The nucleus contains mechanisms which determine which genes are active in the production of appropriate messenger RNA. This is the basis of cellular differentiation. Thus the cells of the pancreas produce the specific enzymes of pancreatic secretion, the activities of the erythrocyte are directed almost exclusively to the synthesis of the protein haemoglobin, and muscle cells produce large amounts of the proteins actin and myosin. All body cells contain the appropriate genetic information for the manufacture of all these proteins but in most body cells this information is not put to use, in other words the particular genes concerned are repressed. That *all* body cells, even when highly differentiated contain all the information for *all* metabolic processes was dramatically illustrated by recent experiments with toad larvae. The intestine was removed from larval toads and the highly differentiated epithelial cells were isolated. The cell nucleus was then removed from an epithelial cell and injected into a toad egg which had been treated with ultra-violet light to inactivate its own nucleus. This egg, containing a nucleus from a fully differentiated and specialized cell of a toad larva, continued to develop normally into a larva and eventually into a normal adult toad.

If all cells produced all the proteins for which genetic information is available then chaos would reign in the differentiated organism. All cells would produce plasma proteins, haemoglobin, contractile proteins, all hormones and so on. Some hint of this state of affairs can be seen in the case of some tumour tissue in which the normal regulation of gene activity is disturbed and genes previously inactive begin to express themselves in terms of protein synthesis. Thus in some tumours of breast or lung tissue genes which are active only in adrenal tissue become derepressed and flood the organism with hormones which drastically disturb the body's regulating mechanisms.

Adaptation

Even in a particular type of differentiated cell the activity of the genetic apparatus is variable. Genetic repression which can be modulated permits a more flexible organization of function in time. This is the basis of adaptation of the organism to fluxes in the internal or external environment. A pre-requisite of this type of adaptation is that genes previously repressed can be derepressed when a situation arises which call for the expression of a particular function.

Thus some species of mammals and birds have restricted periods during the year during which they breed. The timing of this activity is often related to seasonal fluctuations in environmental conditions and food supplies. It is obvious that in periods of the year in which breeding does not occur it would be inappropriate and uneconomical to maintain the structure of the reproductive organs at the level which is attained during the breeding season.

Thus in female mammals during the anoestrous period a variety of retrogressive changes occurs in the reproductive organs. Sexual behaviour is lost, the uterus shrinks in size, there is a thinning of the vaginal epithelium—indeed the vaginal aperture may become sealed, the ovary shrinks, egg production stops and sex hormone production declines. The re-awakening of many of these structures as the breeding season approaches must be associated with the "switching on" of very many genes. In the uterus for example, there is an intensive synthesis of the contractile proteins. The entire energetics of the uterine muscle cell is awakened and energy-rich phosphate compounds accumulate in increasing amounts. The activities of the cell membrane also changes which expresses itself in a rise in the trans-membrane electrical potential.

Some Possible Factors which are Concerned in Regulation of Gene Activity

1. *Nucleohistones.* In the nuclei of most body cells the DNA of the chromosomes is associated with two other types of protein, protamines and histones. These are basic proteins containing a high proportion of the amino acids arginine and lysine. The structure of nuclear DNA is that of a double helix about 2mμ in diameter. The finest chromosomal threads that can be seen with the light microscope are much thicker than these, being 100–200 mμ diameter. It is not clear how the double helices of DNA are organized to form these thicker chromosomal threads but it may involve an aggregation of many DNA double helices or a complex folding of a single long DNA double helix.

The nucleoprotamines and histones appear to be wound round the double helix of DNA lying in the grooves of the helix (Fig. 26).

The function of these basic proteins is not clear. It is possible that they may carry some genetic information but it seems likely that they fulfil other functions. Because they are basic proteins they neutralize the net negative charge of the DNA molecule and this may facilitate the folding or aggregation of DNA molecules, for example to permit close packing of DNA into the head of the sperm. They may also perform other structural functions by acting as bridges between DNA molecules in the structure of the chromosome.

Another possible role of histones which is of special relevance in this discussion is that they may play a part in the differentiation of cells by inactivating part of the genetic material. Thus it may be

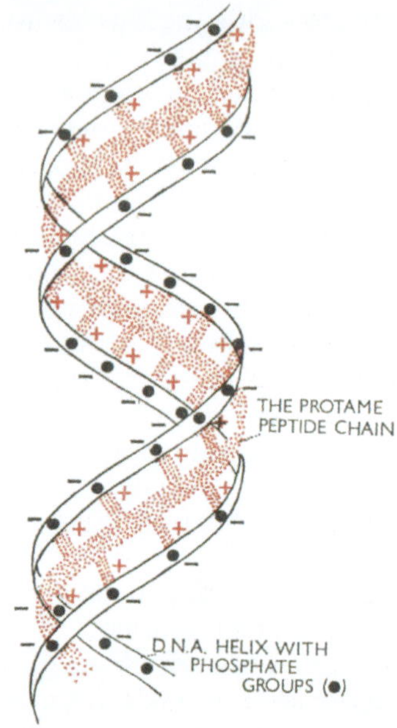

FIG. 26. *Representation of a suggested structure of nucleoprotamines.* The peptide chain of the protamine is supposed to lie in the shallow groove of the double helix of DNA. The positive charges of the arginine side groups are electrostatically attracted to the negatively charged phosphate groups of DNA. The structure of histone is not so clearly established as nucleoprotamines but it seems likely that they also lie in the grooves of the DNA double helix.

---

that during the development of the cell histones could be manufactured and become bound to particular regions of DNA, filling in the grooves of the double helix and making it inactive in the synthesis in messenger RNA. In the differentiated cell large areas of DNA could be structurally and functionally "covered" by histones. This view is supported by some experimental evidence. In some tissues the act of derepression of the genetic material is associated with a loss of histone from the chromatin. Allfrey and Mirsky removed part of the histone from a suspension of thymus cell nuclei by treatment with the proteolytic enzyme trypsin. They found that this was followed by a marked increase in the synthesis of messenger RNA.

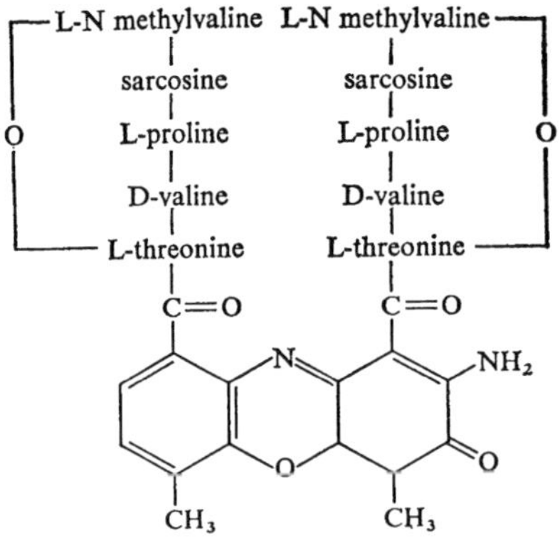

FIG. 27. *The antibiotic actinomycin D.*
This complex substance combines with the base guanine in the DNA molecule. The molecule of actinomycin sticks out of the DNA helix and prevents DNA from acting as a template for the synthesis of RNA.

Thus it seems likely that histones are in some way concerned in the modulation of genetic activity of the cell, and thus in the processes of cellular differentiation and adaptation.

2. *Hormones.* It seems now certain that in many tissues modulation of gene activity is effected by certain hormones. The various lines of evidence which support this view include the following:—

(*i*) An early increase in the rate of nuclear RNA synthesis following administration of the hormone.

(*ii*) The appearance of increased amounts of cytoplasmic RNA in particular that with the characteristics of messenger RNA after treatment with the hormone.

(*iii*) Inhibition of the effects of the hormone by previous treatment with the agents actinomycin D and puromycin (page 68). Actinomycin D (Fig. 27) is a substance which base pairs specifically with guanine as it is found in the double helix of DNA and the complex formed prevents DNA from acting as a template for RNA synthesis. Thus actinomycin blocks the synthesis of messenger RNA.

(*iv*) The appearance of new types of cytoplasmic protein following the administration of the hormones.

Later we shall be considering some of the evidence that hormones such as cortisone, insulin, oestrogens and testosterone may produce some of their effects by acting on the genetic material of the cell.

3. *Enzymic induction and repression.* Some idea of the nature and complexity of gene regulation has been derived from studies of micro organisms, and these may have some bearing on the nature of differentiation and adaptation of cells in higher organisms.

In micro organisms the manufacture of particular enzymes may require in addition to possession of an appropriate gene (the so-called structural gene) the presence of a particular substrate in the environment to act as an inducer. The presence of a particular substrate in the medium in which micro-organisms are cultured will often lead to the appearance of various enzymes concerned in the stages of metabolism of the substrate. This phenomenon is called enzymic induction. The reverse situation also occurs in which the presence of a particular substrate can inhibit the synthesis of a particular enzyme.

From genetic and biochemical studies in micro organisms Jacob and Monod were able to show that the synthesis of a particular enzyme by a micro organism is determined not only by a gene which specifies the amino acid sequence of the enzyme in terms of messenger RNA (the structural gene) but also by a hierarchy of other genes. One such gene is the regulator gene which may control the activity of a group of distant structural genes by means of a repressor substance. The repressor substance produced by the regulator gene does not seem to act directly on the structural genes. Instead it acts by way of another gene which is located adjacent to the group of structural genes controlled by the regulator (Fig. 28). To this gene is given the name of operator. The intracellular repressor produced by the regulator acts on the operator to switch off the activity of the whole groups of structural genes.

Jacob and Monod suggested that the repressor substances produced by regulator genes are protein in nature and contain two active sites, one capable of binding with the operator gene and one specific to the inducer substance. They suggested that on introduction of the inducer substance into the medium this combines with the repressor protein and this results in such a modification of the structure of the repressor molecule that it can no longer combine with the operator gene. The activity of the operator gene is now released from inhibition and it then activates the structural genes.

When Jacob and Monod put forward this scheme of regulation of metabolic activity in bacteria there was no direct evidence of

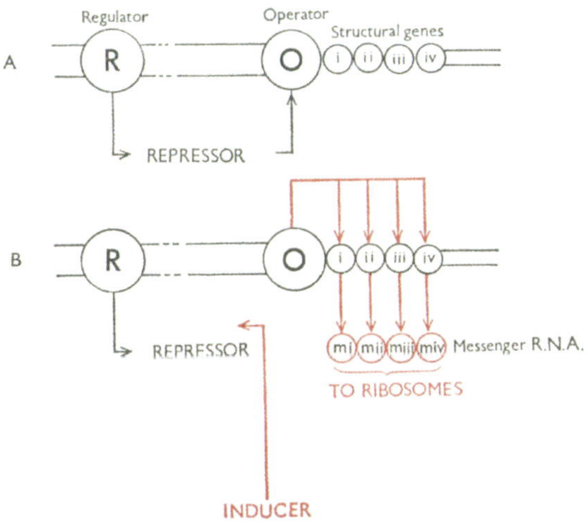

FIG. 28. *Scheme to illustrate Jacob and Monod's theory of genetic regulation.*

A. Shows four regulator genes (i–iv) and the adjacent operator gene (O). These genes are inactive, i.e. repressed, because of the activity of a regulator gene (R) which produces a repressor substance inhibiting the activity of operator gene.

B. Shows the effects of adding an inducer substance to the medium in which the bacteria are living. The inducer substance combines with and inactivates repressor substance. The operator gene is now released from inhibition and it activates the structural genes i–iv. These now produce specific messenger RNA, mi–iv, which passes to the ribosomes to instruct the synthesis of four distinct enzymes.

Note that although the repressor substance is shown arising from the regulator gene this protein, like all cell proteins, is coded for by a segment of DNA which produces a specific messenger RNA instructing the synthesis of the repressor molecules on the ribosomes.

---

the nature or indeed even of the presence of the repressor substance. In 1966 studies by American teams of scientists provided this direct evidence. The repressor has been isolated for the metabolic system which first led Jacob and Monod to put forward their concept—that is the repressor substance which controls the genes determining the enzymes which metabolize lactose in the bacterium E. coli. The repressor does indeed bind itself to a section of DNA. In 1968 an account was published of work which identified the region of the genetic material of E. coli where the synthesis of messenger RNA for lactose splitting enzymes begins. When a stretch of DNA is copied in terms of messenger RNA an enzyme called RNA polymerase appears to become attached at a particular point on the DNA chain and works along like a zip fastener, attaching itself to successive stretches of DNA and producing messenger RNA as it moves along. It seems that the operator gene lies between the point of attachment of the RNA polymerase and the structural genes which code for messenger RNA. This discovery points to a straightforward way in which the repressor substance can prevent the copying of the structural genes. When the repressor substance attaches itself to the operator it acts like a thread caught in a zip fastener to prevent the progress of R.N.A. polymerase along the segment of DNA.

Inducible and repressor enzymes have now been described in animal cells. In general the response of animal cells to changes in substrate composition are slower and smaller than those of bacteria. The system described in bacteria may provide some pointers to the way in which hormones can alter gene activity, e.g. by inactivation of an intracellular repressor or inducer.

The very brilliance of this concept by Jacob and Monod has for a few years somewhat blinded us to the possibility that other mechanisms may also be involved in the changes in protein synthesis which some hormones produce. This possibility will be examined in more detail in Chapter 9 in which evidence will be presented that growth hormone can affect not only messenger RNA formation but also the transcription of this message by the ribosomes.

Rather different mechanisms which regulate gene activity are indicated by the work of Professor Harris of Oxford. In 1965 he described experiments in which he introduced the nucleus of a cell of one species of animal into the cell of a different species. The donor cells were bird erythrocytes (red blood cells) which are so highly specialized that virtually all of the genes are switched off. These cells produce no DNA and only very small amounts of RNA. Harris introduced the nuclei of bird erythrocytes into human cells which were active in the synthesis of both DNA and RNA. He found that after introduction into the human cell the bird erythrocyte nucleus increased several times in size, its highly condensed chromatin became opened up and it became active in RNA synthesis. He argued that these results were not compatible with Jacob and Monod's repressor hypothesis. The results indeed suggested an *activation* of the DNA of bird erythrocyte nuclei by a cytoplasmic factor(s) from the recipient cell. It is suggested that the activator causes nuclear swelling and opening up of the dense inactive chromatin so permitting the DNA to be copied by RNA polymerase. The amounts of RNA produced by the erythrocyte nuclei were indeed proportional to the degree of nuclear enlargement. These results have been confirmed in studies of the effect of incubation of erythrocyte nuclei with mouse liver cell cytoplasm.

It would appear that there may be a variety of mechanisms involved in the regulation of the activity of DNA and it would be premature to speculate on the significance of these various findings for the action of hormones on DNA.

The Evidence that Cortisone Acts on the Chromatin of the Cell

We are concerned here with the effect of adreno-cortical hormones on carbohydrate metabolism in the liver. Glucose and glycogen can be manufactured in the liver from non carbohydrate sources, such as some amino acids, glycerol, pyruvic acid, lactic acid by a process called gluconeogenesis. The generation of glucose from amino acids is a process of some complexity and involves many intermediate steps between the initial substrate and the end product (glucose). At an early stage in the metabolism of these amino acids there is a removal of the α-amino group, a process called deamination.

The supply of glucose from non carbohydrate sources varies according to the requirements of the organism. It is maximal during fasting when there is no exogenous supply of carbohydrate. The activation of gluconeogenesis requires the presence of a functional adrenol cortex. The complexity of the transformation of amino acid to glucose is such that the activation of no single enzyme system would automatically result in an increased rate of generation of glucose. In fact it has been known for many years that following the injection of cortisone into mammals there is a marked increase in the amounts of several enzymes in liver cells. These enzymes include glutamic-tyrosine transaminase, glutamic-alanine transaminase—enzymes which catalyse the initial transfer of the α-amino group of the amino acids, and others such as trytophan pyrolase, aldolase, fructose diphosphate phosphatase and glucose-6-phosphatase.

In 1962 it was found that the administration of cortisone to rats caused the appearance of *new* enzyme in liver cells. This was shown by giving to the animals amino acids which were labelled with radioactive atoms together with the cortisone. The rats were later sacrificed and an enzyme, glutamic tyrosine transaminase, was isolated from liver tissue. It was found that the isolated enzyme was radioactive indicating that it had been synthesized from amino acids under the influence of cortisone.

The effects of cortisone on the liver can be prevented if the animals are also given a substance puromycin. Puromycin has a chemical structure which partially mimics the action of that cytoplasmic RNA known as transfer RNA which is responsible for "gathering up" appropriate amino acids in the cytoplasm and bringing them into close relationship to the surface of ribosomes where they are utilized in the synthesis of protein molecules under instruction by molecules of messenger RNA (page 94). Puromycin prevents the normal growth of the peptide chain on the ribosomes and thus it inhibits the synthesis of protein in a very specific manner. This effect of puromycin in preventing the response of liver cells to cortisone is further evidence that the increased amount of enzymes present in liver cells after treatment with cortisone is due to the synthesis of new protein.

In addition to an increase in the amounts of various enzymes in liver cells, cortisone treatment also results in an early increase in the rate of nuclear synthesis of RNA which is followed by the appearance of increased amounts of RNA in the cytoplasm of the cell, including that RNA with the characters of messenger RNA. If one could prevent this effect of cortisone then one could also prevent the increased synthesis of enzymes since RNA is indispensable for protein synthesis. One can indeed prevent the action of cortisone on nuclear synthesis of RNA. The substance actinomycin (page 63) base pairs specifically with guanine in the double helix of DNA. The molecule of actinomycin sticks out of the DNA helix—as Bonner graphically describes it, "like a banderilla out of a bull"—and prevents DNA from acting as a template for the synthesis of RNA.

Although cortisone undoubtedly has the above mentioned effect on the synthesis of liver enzymes the significance of this effect for the regulation of blood glucose is at present still a subject of debate. Figure 29 shows the results of an experiment using rats, to test the effect of actinomycin on the changes which are produced by hydrocortisone on liver glycogen and blood glucose. Four groups of rats were taken and food was withheld from them for twelve hours. Groups:—

(*i*) Starved control rats. No treatment.

(*ii*) Starved and treated with actinomycin D.

(*iii*) Starved and treated with actinomycin D and hydrocortisone.

(*iv*) Starved and treated with hydrocortisone.

The bar graphs show the glucose concentration in blood (mg %), liver glycogen content (g %) and the activity of a liver enzyme phosphoenolpyruvate carboxykinase (PEP) in each group of animals.

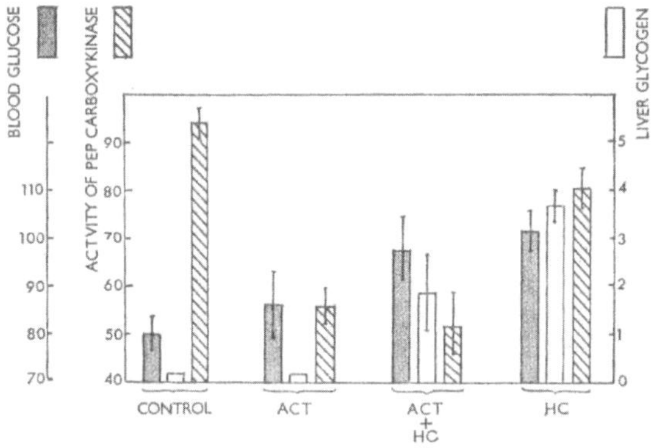

FIG. 29. *The effect of actinomycin on the responses of the rat to hydrocortisone* (*after Ray* et al. *1967*).

The enzyme PEP is only one of many involved in gluconeogenesis, but it was chosen because it is a key catalyst in gluconeogenesis. The results show that in the normal animal fasting produces an increase of PEP carboxykinase from 50 units (in the normal fed animal) to 90 units after fasting for twelve hours. Treatment with actinomycin prevented the response of PEP carboxykinase *but* did not prevent the animal from maintaining a normal glucose concentration in blood. Treatment with hydrocortisone, as expected, increased the amounts of glucose in blood and glycogen in liver, and increased the activity of PEP carboxykinase. The addition of actinomycin to hydrocortisone prevented increase of

PEP carboxykinase above the normal level (50) *but* did not significantly reduce the influence of hydrocortisone on maintenance of blood glucose, although it did reduce the response of liver glycogen.

These results show clearly that in short term experiments the ability of hydrocortisone to increase the glucose content of blood is not significantly reduced if the action of the hormone on RNA synthesis (and thus enzyme synthesis) is blocked by actinomycin. In experiments repeated on rats from which the adrenal glands had been removed similar results were obtained except that in this situation the effect of hydrocortisone on liver glycogen production was reduced by 60% when actinomycin was also administered.

One must conclude that cortical hormones must have other actions in addition to their effects on messenger RNA production in the liver, which promote the process of gluconeogenesis. One such action is to increase the availability of amino acids to the enzyme systems in the liver—that is to increase the supply of raw material for protein synthesis. Cortical hormones affect protein metabolism in many body tissues, reducing the rate at which amino acids are incorporated into protein. These amino acids now become available for gluconeogenesis in the liver.

Summary. Thus two mechansisms of action of adrenocortical hormones at the molecular level must be considered.

1. The hormones release gluconeogenic precursors (e.g. amino acids) from peripheral tissues which become available for the liver for glucose synthesis.
2. The hormones stimulate the hepatic synthesis of enzymes concerned with gluconeogenesis. This effect does not seem to start immediately and the effects on carbohydrate synthesis are delayed somewhat.

In the present state of our understanding both mechanisms are involved although the primary effect would seem to be on the peripheral tissues. However it seems safe to say that induction of liver enzymes is an integral part of a maximal response to adrenal corticoids.

CHAPTER 7

Hormones Acting at the Level of the Gene: II Sex Steroids—Oestrogens and Androgens

The primary organs of reproduction—that is the ovary and testis—serve two functions. First, they produce the germ cells, the ova and sperm. Second, they produce hormones which regulate the behaviour of the animal and the structure and function of a host of different tissues, so that the metabolism of the organism is tuned to mating, fertilization of the ovum, and the nurture of the fertilized ovum in the reproductive tract of the female. We shall examine briefly the diverse effects of sex hormones and their possible mechanisms of action.

Oestrogens

The term oestrogen is applied to a substance which produces a cornification of the vaginal epithelium of the mouse similar to that which occurs during the phase of oestrus. A variety of substances will produce this effect. Some are naturally occurring compounds such as oestrone or oestradiol (Fig. 30) and others, such as

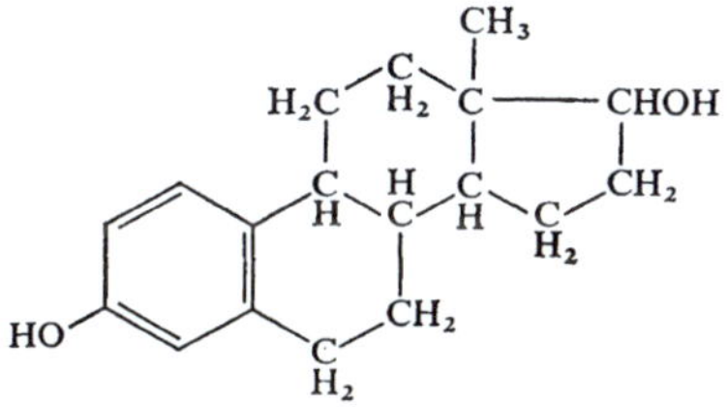

FIG. 30. *The structure of oestradiol.*

stilboestrol are synthetic substances which do not normally occur in mammals. These various substances are known collectively as oestrogens. They may differ considerably in their potency. Oestradiol is the most potent naturally occurring oestrogen.

Sources of Oestrogen. Under normal conditions the ovary is the chief source of oestrogen. Several components of the ovary appear to be able to produce the hormone although the outer wall of the

Graafian follicle in the ovary (Fig. 31) is probably the most

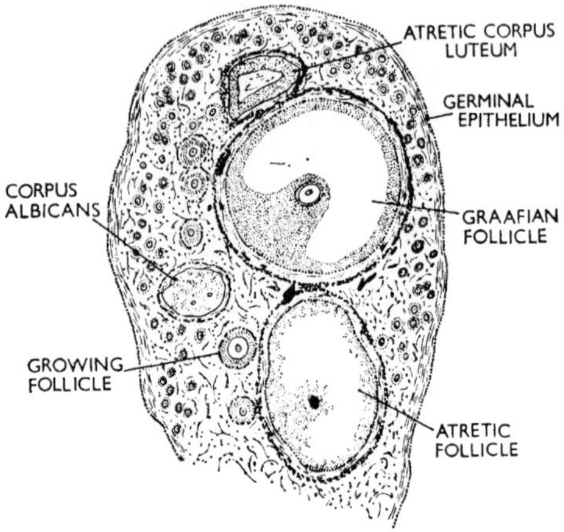

FIG. 31. *Drawing of a section of a mammalian ovary showing mature and immature Graafian follicles.*

important source. Other tissues may also produce oestrogen. The adrenal cortex may produce small amounts, and a much richer source is the placenta which produces large amounts of the hormone. Moreover even the male produces some oestrogen. In many species the quantities produced are small and the significance of this hormone in the male is poorly understood. However, in at least one mammalian species the male produces large amounts of oestrogen. The urine of the stallion contains amounts of oestrogen (as oestrone) which sometimes exceeds that of the pregnant mare.

Actions of Oestrogen. Although the oestrogens are defined in terms of their action on the epithelium of the vaginal canal their sphere of action is infinitely greater than this. To describe fully the action of oestrogens on the host of tissues which respond to the hormones would need a volume far larger than this one.

1. *Reproductive tract.* The reproductive tract of the female mammal, including both internal and external genitalia, is dependent upon a continuous supply of oestrogen for its growth and maintenance. The administration of oestrogen to a sexually immature female is followed by dramatic changes in the growth and function of the various tissues. In the vagina these include a thickening and cornification of the epithelium, an increase

in blood supply, growth and accumulation of fluid in the muscular and connective tissues of the vaginal wall, the appearance of glycogen in the cells of the vaginal epithelium and changes in the electrical potential that we can record across the vaginal wall.

The uterus and Fallopian tubes are also dependent upon a supply of oestrogen for growth and maintenance. Oestrogen affects both the muscular wall and the mucous membrane lining these organs. The effects of the hormone on the muscle cells is considered in more detail below (page 78).

2. *Mammary glands.* The secretion of milk by the mammary glands occurs very soon after birth of the young. The growth and function of these glands is thus closely linked to other aspects of reproductive function. A variety of endocrine mechanisms are involved in this synchronization of function. In fact the regulation of the mammary gland, arising as it did late in the phylogeny of vertebrates is more complex than the regulation of the reproductive organs themselves and is still incompletely understood.

The effects of administration of oestrogen to animals varies considerably from one species to another making it difficult to assess the precise role of oestrogen in breast development and function. Certainly the hormone is indispensable for breast development and the hormone cooperates with other hormones, e.g. progesterone (secreted by the ovary and placenta) and prolactin (secreted by the anterior pituitary gland) in regulating breast function. In some animals, e.g. guinea pigs and monkeys, oestrogen treatment alone will stimulate a development of mammary gland structure similar to that which is achieved during pregnancy.

3. *The hypothalamus and pituitary gland.* The activity of the ovary (and testis) is regulated by endocrine secretions of the anterior lobe of the pituitary gland (Fig. 4). These secretions are called gonadotrophins. In the absence of the anterior lobe of the pituitary the gonads remain infantile in structure and function and neither germs cells nor sex hormones are produced.

The relationship between the pituitary gland and gonad (ovary or testis) is however a reciprocal one and endocrine secretions of the gonads act back upon the pituitary to modify the secretion of gonadotrophins—i.e. a feed back mechanism.

It is probable that the sex hormones do not act on the pituitary gland directly, or at least not in quantities which are found in the normal animal. Instead the sex hormones act at a higher level of control, i.e. on the hypothalamus (page 10). The hypothalamus regulates anterior lobe function by means of chemical substances —local hormones if you like—which are transported from the hypothalamus to the pituitary in the portal system of blood vessels which connects the two regions. The simplest explanation of this feed back mechanism is that some neurones in the hypothalamus are directly sensitive to hormones in the cirulating blood. If a small amount of oestrogen is implanted by way of a needle into a suitable region of the hypothalamus it produces an effect on the secretion of gonadotrophins by the anterior pituitary gland which is 125 × greater than if the hormone was injected under the skin of the animal. This supports the view that the hormone acts directly on the hypothalamus. Not all areas of the hypothalamus are sensitive to sex steroids but some regions are exquisitely so. The distribution of cells reacting to oestrogen has been studied by various techniques. The technique of autoradiography (page 27) for example has shown that after the injection of radioactive oestrogen into the animal certain tissues accumulate and retain the hormone to a degree. These target tissues include certain areas of the hypothalamus.

In addition sex hormones also influence those areas of the hypothalamus which are associated with sexual behaviour. Thus the administration of oestrogen to a female cat which is not in breeding condition will rapidly be followed by the appearance of breeding behaviour. This effect of oestrogen is seen even in the absence of the pituitary gland but is lost after localized destruction of part of the anterior hypothalamus.

4. *Other effects of oestrogen.* In addition to the effects on the above structures oestrogen also has marked influences on body growth and body weight and on individual tissues such as skin, adrenal gland, and on the metabolism of water and electrolytes. In the hen oestrogen has a distinct effect on the liver and causes liver cells to produce the proteins which will appear in the egg. Although the cock bird does not normally manufacture such proteins even he will respond to oestrogen by the synthesis of large amounts of yolk proteins by the liver.

Androgens

These are substances which promote the development of male characters, both structural and behavioural. Thus when injected into the capon (a castrated cockerel) androgens cause marked

FIG. 32. *The effect of painting comb and wattles of the capon with a solution of testosterone.*
A. Untreated capon. Surface area of wattles about 10 cm^2.
B. Capon treated with testosterone for two weeks. Note increase in size and vascularization of comb and wattles. Surface area of wattles about 65 cm^2.

---

development of comb and wattles, and the appearance of aggressive behaviour (Fig. 32). The most potent naturally occurring androgen is testosterone (Fig. 33), a steroid sex hormone.

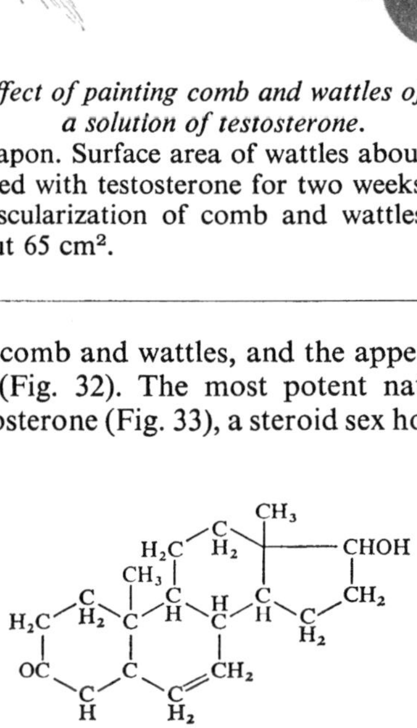

FIG. 33. *The structure of testosterone.*

The main source of androgen in the male is the testis. The testis contains two types of tissue; the bulk of the organ is composed of the seminiferous tubules which produce the sperm, and in between these tubules is the interstitial tissue, the source of testosterone (Fig. 34).

Actions of Androgen. The actions of androgen in the body are widespread and they affect many processes which are not directly related to reproduction. Further, androgens can produce in the female organism effects which are similar to those in the male—

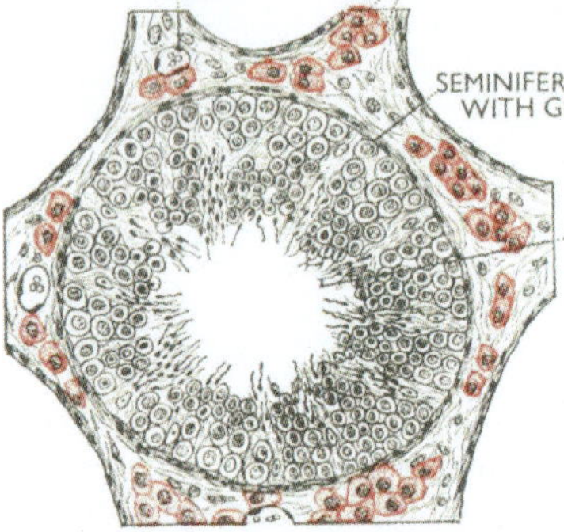

FIG. 34. *Sketch of part of a section of the testis showing the interstitial cells lying between the seminiferous tubules.*

with obvious exceptions where the female does not possess appropriate tissues to respond to the hormone.

Genital tract. Each region of the genital tract and its associated glands is regulated by testicular androgen. The administration of androgen to the sexually immature or castrated male produces marked effects on the scrotum, seminal vesicles, prostate gland, penis, bulbo-urethral glands (Fig. 35). If the hormone is given to pregnant females it disturbs the development of the genital tract of female foetuses, producing a masculinized genital tract. Even in the fully differentiated female androgens may produce some features of masculinization, e.g. hypertrophy of the clitoris which is the female homologue of the male penis.

Non-genital structures. Androgens also influence non-genital sex characters such as hair distribution, voice, fat distribution, muscle size, etc. Like oestrogen, testosterone influences the hypothalamus and thus indirectly the anterior pituitary gland.

The hypothalamus is influenced in other ways by testosterone. In the rat and other mammals the hypothalamus of the newly born animal is apparently sexually indeterminate and is capable of differentiating into either male or female type irrespective of the sex of the gonads, provided that appropriate conditions are met. The female type of hypothalamus causes a rhythmical secretion of gonadotrophins by the anterior pituitary gland so that the ovaries undergo the cyclic processes of follicular development and corpus luteum formation. The male type of hypothalamus determines a relatively steady secretion of gonadotrophins during the breeding season. The main factor which determines the direction of differentiation of hypothalamic function into male or female type is the presence or absence of androgen. In the absence of male hormone (i.e. in a normal female or in a male castrated at birth) the hypothalamus differentiates into the female type. Thus if if ovaries are later transplanted into a male castrated at birth the ovaries will undergo normal cyclic activity under the influence of rhythmical secretion of gonadotrophins. If male sex hormone is administered early in life the hypothalamus differentiates into the male type, irrespective of the sex of the animal. In addition to these effects androgen is also concerned in determining the direction and intensity of male sexual drive.

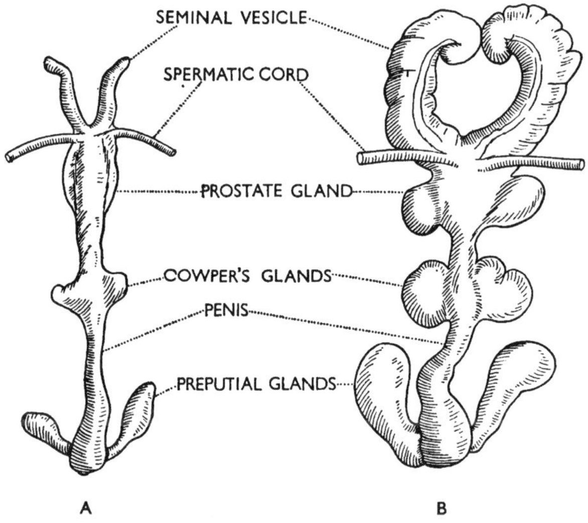

FIG. 35. *The effect of testosterone on the reproductive tract of a castrated male rat.*

A. Genital tract of untreated castrate.

B. Genital tract of castrate after a single injection of testosterone (2 mg) eleven days previously.

A variety of other tissues are also influenced by androgen including the kidney, adrenal gland, liver, thyroid, thymus, skin (pigmentation, blood flow and secretions) and bone marrow. It is obvious that very many body tissues are capable of responding to androgens as they do to oestrogens. Some tissues have developed this responsiveness to a degree in which major changes in tissue activity result from the application of the hormone in physiological quantities.

The Mode of Action of Sex Hormones

Oestrogen. We have seen that the consequences of administering oestrogen or androgen to an animal are varied and that many different tissues are involved in the response to the hormone. Even in a single cell, e.g. the smooth muscle cells in the wall of the uterus, the effects of oestrogen may be multiple. Thus after the administration of oestrogen the uterine muscle cells increase in size, they synthesize more of the contractile proteins, and accumulate energy rich phosphate compounds. There are also changes in the enzyme constitution and activity and thus in the respiratory rate. The electrical potential that can be recorded across the smooth muscle cell membrane increases and there is a greater tendency for some cells to generate action potentials spontaneously—i.e. in the absence of nervous or chemical stimulation. These spontaneously appearing fluxes in the electrical properties of individual smooth muscle cells are conducted over large stretches of the muscle and are followed by waves of contraction spreading over the uterus. If the uterus from such an animal is removed and placed in a bath of warm oxygênated physiological saline solution and connected to a lever we can readily record the spontaneous activity which is independent of nervous influence (Fig. 36). This spontaneously active organ of the oestrogen treated animal is also very sensitive to chemical stimulants such as oxytocin.

One of the first observable effects of oestrogen on the uterine cells is a dramatic increase in the rate of synthesis of RNA. This occurs in less than thirty minutes after administration of the hormone. Initially there is a rise in the amount of messenger and transfer RNA which is followed by the appearance of increased ribosomal RNA. But even earlier than the appearance of RNA is the activation of an enzyme RNA—DNA polymerase which is involved in transcribing the message of lengths of DNA into messenger RNA.

These effects on RNA synthesis ultimately lead to an increase in the various cellular proteins. This sequence of events is what we would expect if oestrogen triggered off the changes in the cell by a primary action on the gene. Indeed if actinomycin is administered to the animal before the oestrogen then none of these effects on protein synthesis appear. If all the many effects of oestrogen in uterine cells are to be explained in a similar way then we must postulate that oestrogen is activating a considerable number of genes. The problem remains of explaining how a single hormone, oestrogen, can activate a whole range of genes in a special

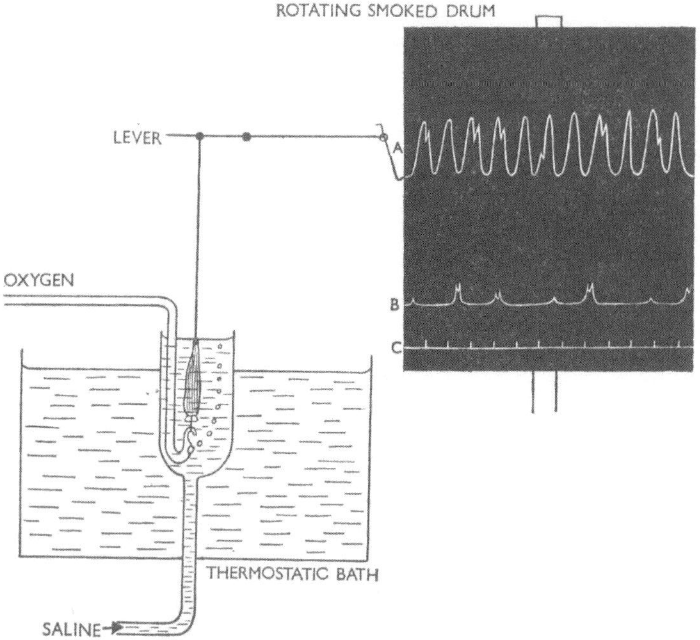

FIG. 36. *Figure showing a simple form of apparatus for recording the activity of isolated organs.*

The uterus is mounted in a chamber of physiological saline solution (inlet for fresh saline below), gassed with oxygen and its upper end connected to a lever. The lower end of the uterus is fixed to the oxygen supply tube with thread. The activity of the uterus is recorded by friction of the mobile lever tip with the surface of a revolving smoked drum. The record A shows regular contractions of near maximal amplitude of a uterus from a rat in oestrus (i.e., under the influence of oestrogen). B shows irregular low amplitude contractions shown by a uterus from a rat not in oestrus. C indicates time trace in minutes.

sequence and to different degrees. And of course the possibility exists that the hormone has some effects on the cells which are independent of direct gene activation.

In different target organs a different set of genes will be activated. Thus in the liver of the hen the activated genes include those carrying the information for the construction of the egg yolk proteins phosphvitin and lipovitellin. The protein phosphvitin contains large amounts of the amino acid serine and prominent among the types of transfer RNA which appear in the hen liver cells after oestrogen treatment is that transfer RNA which is involved in the inclusion of the amino acid serine into the growing protein molecule on the ribosomes.

Androgen. The treatment of animals with androgens produces, like oestrogen, marked changes in the structure of target tissues—in their size, blood flow, secretions, cytological structure and biochemical composition. Recent evidence suggests that, like oestrogen, some of the effects of testosterone appear to be due to gene activation.

It was found in 1962 that ribosomes from the prostate gland of the castrated rat synthesized protein at a higher rate when the animal had received testosterone. Since the addition of hormone to the ribosomes from a castrate did not also stimulate protein synthesis it seems likely that the hormone acts at a stage preceding ribosomal function. It was found later that within an hour of administering testosterone to castrate rats that there was an increase in RNA synthesis in the prostate gland. In one species administered testosterone has been found to be bound to nuclear DNA (duck's preen gland).

On bone marrow testosterone has another effect. The action of testosterone here is to stimulate the production of red blood cells, which involves multiplication of red cell precursors in the bone marrow. Autoradiographic studies of bone marrow incubated with testosterone and a radioactive precursor of DNA (tritiated thymidine) have shown that the hormone stimulates the formation and replication of DNA.

CHAPTER 8

Hormones Acting at the Level of the Gene: III Ecdysone

The Life Cycle of Insects

The development of insects such as butterflies, house-fly, midges, fruit flies, etc. occurs as a series of transformations from the egg to the adult organism (Fig. 37). From the egg of higher insects hatches an organism, the larva, which is quite different in structure and behaviour from the adult organism. Inside the larva are dormant rudiments of the imago. Increase in size of the larva is not a continuous process, but occurs only at restricted periods in the life of the larva when the skin is shed (i.e. at moulting). At the last moult in the life of the larva the organism is transformed into a hard skinned relatively immobile pupa in which the rudimentary organs of the imago begin to develop at the expense of larval tissues. At the last moult in the life of the organism the pupal skin is shed to reveal the fully developed imago.

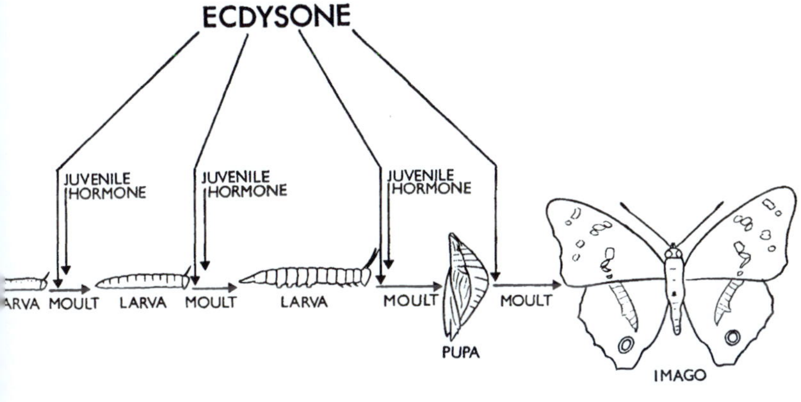

FIG. 37. *Life cycle of an insect* (Lepidoptera—*Purple Emperor*).

The moulting of insects is regulated by several endocrine tissues including neurosecretory cells in the brain, the corpora allata and the prothoracic glands. Ecdysone, the so-called "moulting and metamorphosis" hormone is produced by the prothoracic glands after these have been activated by a hormone from the brain. This hormone controls the events of the moult. Another hormone produced by the corpora allata—called juvenile hormone—determines the kind of moult with is initiated by ecdysone. When juvenile hormone is present the larval characters are maintained and the moult is larva → larva. In the absence of juvenile hormone the moult is either larva → pupa or pupa → imago.

The Mode of Action of Ecdysone

Polytene Chromosomes. The mode of action of ecdysone has been studied in Dipteran larvae, e.g. Drosophila (the fruit fly) and Chironomus (midge). These larvae have a cytological peculiarity in certain tissues which has permitted the locus of action of ecdysone to be demonstrated in a dramatically direct fashion. This cytological peculiarity is the presence of giant or polytene chromosomes in salivary gland and other tissues. These structures in the nuclei of salivary gland cells of certain Dipteran larvae were recognized late in the 19th century. It was not until 1923 that these structures were recognized as being equivalent to the chromosomes of typical animal cells. In typical body cells chromosomes as distinct small dense bodies are visible with the light microscope only during the process of cell division. During the resting phase of the cell—interphase—the chromosomes are no longer visible since the chromatin material of which they are composed is in the form of long fine threads which have only recently been resolved using the electron microscope. The polytene chromosomes of larval Diptera are so much larger than typical chromosomes that they are readily visible during interphase with the light microscope (Fig. 38). When they are fully grown they are about 100 × thicker and 10 × longer than normal chromosomes. In these tissues which possess polytene chromosomes, growth is by an increase in the size of individual cells rather than by an increase in the number of cells. As each cell grows in size it seems that increasing amounts of chromatin are needed and the chromosomes undergo repeated division and grow in length. When the chromatin of the polytene chromosome replicates itself there is no separation of the genetic material as in typical cells where half of the chromatin enters each daughter cell. Instead the chromosome increases in thickness, and the fully grown polytene chromosomes may contain as many as 10–20,000 single strands.

FIG. 38. *Sketch of the salivary gland chromosomes of* chirinomus thumii *showing the pattern of banding*. (Modified from Fig. 11 in *Animal Cytology and Evolution* by M. J. D. White, Cambridge University Press, 1948.)

The polytene chromosomes show a series of distinct transverse bands which contain DNA and histone (page 61) in high concentration. The paler areas between the darkly staining bands have only small amounts of hereditary material.

Each chromosome has its special pattern of banding, and this pattern is shown by all the homologous chromosomes of the same individual, in fact in all individuals of the same species. There is strong evidence that the bands on the chromosomes represent the Mendelian genes, or at least small groups of genes.

Variability of Gene Action. Although the sequence of the transverse bands on the polytene chromosomes is constant their appearance and activity is not constant. The transverse bands from time to time may undergo what is known as puffing. In puffing the dense transverse band of the chromosome begins to lose its clear outline and presents a diffuse appearance. This may progress to a gross enlargement of the region, producing a thick ring along the axis of the chromosome—a Balbiani ring (Fig. 41). In unpuffed regions of the chromosome the very many fibrils (10–20,000)—called chromomeres—are twisted around one another to form a dense cable-like structure. During puffing the fibrils appear to untwist and pop out of the cable—like an uncoiled watch spring (Fig. 39).

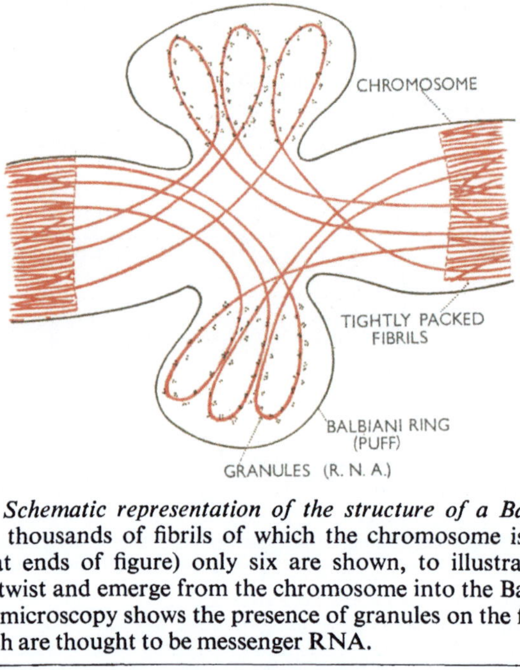

FIG. 39. *Schematic representation of the structure of a Balbiani ring.* Of the thousands of fibrils of which the chromosome is composed (shown at ends of figure) only six are shown, to illustrate how the fibrils untwist and emerge from the chromosome into the Balbiani ring. Electron microscopy shows the presence of granules on the fibrils in the puff which are thought to be messenger RNA.

The puffed areas of the chromosomes contain quantities of RNA. In contrast the unpuffed regions are composed principally of DNA and histone. The puffs are regarded as derepressed genes actively engaged in the synthesis of messenger RNA. Various lines of evidence support this view. Treatment of animals with actinomycin D, which blocks DNA dependent synthesis of RNA, will prevent puffing. One can also detect sites of RNA synthesis using radioisotope techniques. Claus Pelling injected into midge larvae the base uracil (page 56) radioactively labelled with tritium (H_3). This base is used in the synthesis of RNA but not DNA. At various intervals after injection of radioactive uracil, the larvae were killed and cells containing polytene chromosomes removed and placed in contact with an appropriate photographic emulsion. Radiation from tritium darkened the plate. After a suitable interval the plates were processed and showed dark spots corresponding to parts of the cell containing radioactive material. This technique is called autoradiography (page 28). When the larvae were killed shortly after the injection the only components of the cell which were radioactive were the chromosomal puffs and nucleoli. If either actinomycin D or an enzyme which destroys RNA were administered to the larvae before the injection of radioactive uracil then the puffs did not accumulate radioactivity. These experiments clearly indicate that the chromosome puffs are regions of new RNA synthesis. This new RNA is in fact messenger RNA (see below).

Ecdysone and Puffing. Some of the puffs which appear on the giant chromosomes of Dipteran larvae bear no relation to the moulting process and are concerned with other metabolic activities. During each moult however there is a characteristic sequence of puffing of chromosome bands, different puffs appearing at different phases of the moulting process. The same sequence of puffing also follows the injection of a minute quantity of ecdysone into a larva which is not preparing to moult. The response to ecdysone is rapid and puffing at one locus occurs within thirty minutes of injection of ecdysone. The hormone was first extracted from the pupae of silkworm—three tons of silkworm pupae were needed to extract 100 mg of pure hormone. It is a steroid hormone (Fig. 40).

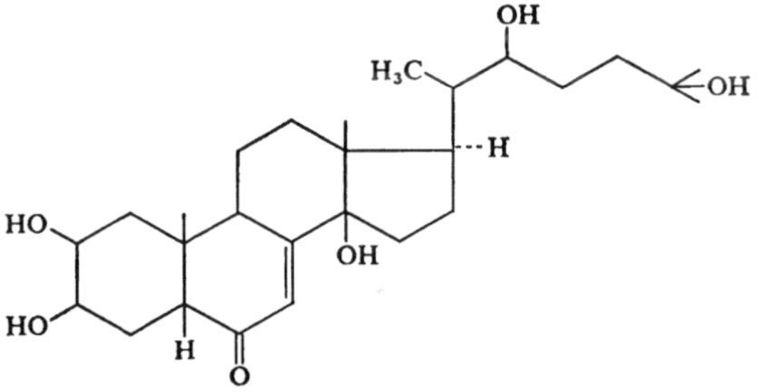

FIG. 40. *The structure of ecdysone (after Karlson, 1967).*

The various puffs which develop under the influence of ecdysone appear at different times after injection of the hormone. Moreover different puffs have different thresholds, that is different amounts of ecdysone are required to initiate puffing in different regions of the chromosomes. Ecdysone not only initiates moulting but its continued presence is needed for completion of the process. If the source of ecdysone—the prothoracic glands—is removed during moulting then chromosome puffing and moulting are halted. Thus if in a Dipteran larva about to moult a ligature is tied round the head end to exclude the prothoracic glands from the rest of the larva then chromosome puffing and moulting do not occur in the region of the larva behind the prothoracic glands.

We can conclude that the hormone ecdysone controls the process of moulting by the derepression of certain genes. Gene activation expresses itself by synthesis of messenger RNA leading to protein synthesis by the ribosomes and the establishment of the metabolic changes associated with moulting (Fig. 41). The appearance of a new kind of messenger RNA following treatment with ecdysone was shown in an elegant fashion by Sekaris and Lang (1964). In the last larval moult which produces a pupa there is a hardening and darkening of the pupal skin. During this moult a hardening agent—N-acetyl dihydroxyphenethylamine—appears in the epidermal cells of the larva. This substance is derived by the decarboxylation of dihydroxyphenylalanine (DOPA) under the influence of an enzyme DOPA decarboxylase. Sekaris and Lang were able to show that during the last larval moult there appears in the epidermal cells of the larva a new kind of messenger RNA which instructs the ribosomes in the production of the enzyme DOPA decarboxylase. They first isolated nuclei from epidermal cells of blowfly larvae after these had been injected with ecdysone. RNA was then isolated from these nuclei and added to a suspension containing ribosomes and the various factors known to be necessary for protein synthesis with the *exception of messenger RNA*. The addition of RNA of blowfly epidermal cells to this system resulted in the synthesis of the enzyme DOPA decarboxylase. Thus the RNA from the larvae has the features of a particular messenger RNA. RNA extracted from control blowfly larvae which had not been treated with ecdysone produced no detectable DOPA decarboxylase by the ribosomal system.

The messenger RNA which directs ribosomal synthesis of the enzyme DOPA decarboxylase is only one type of messenger

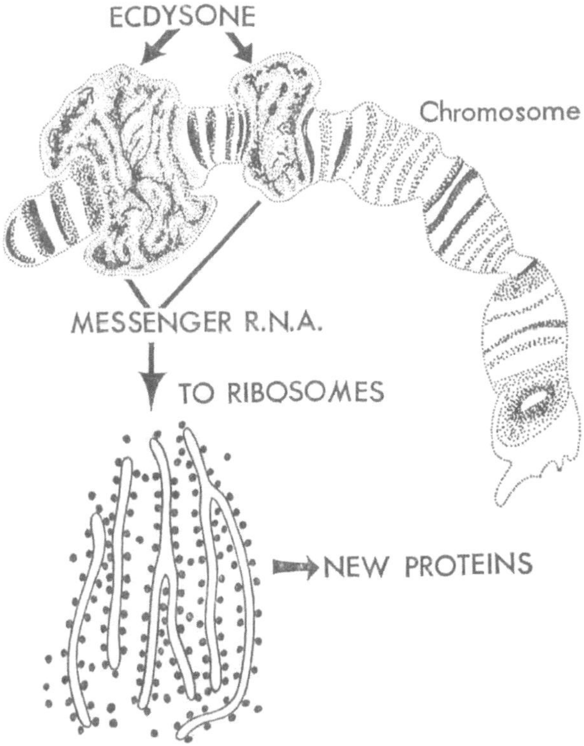

FIG. 41. *The mode of action of ecdysone.*
The ecdysone has activated two areas of the chromosome. The chromosome was drawn from a specimen of salivary chromosomes of a Dipteran larva stained to distinguish DNA (black) and RNA (red). Two Balbiani rings (large puffs) are shown containing considerable amounts of RNA. Messenger RNA leaves these regions of the chromosome and passes out into the cytoplasm of the cell where it directs the synthesis of new proteins on the ribosomes.

---

produced under the influence of ecdysone. Presumably each puff produces a different class of messenger RNA. The amount of ecdysone necessary to produce puffing varies from one region of the chromosome to another. The amounts are, however, small. In Chironomus larvae, puffing is produced at one locus when only 0·000001 microgram per milligram of larval weight is administered.

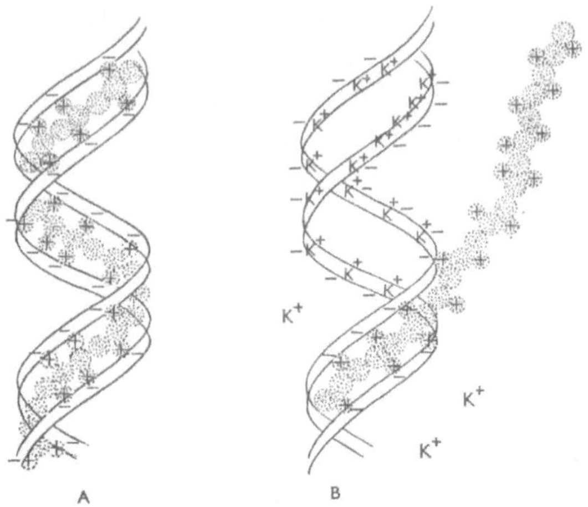

FIG. 42. *Representation of a suggested mode of action of ecdysone*
In A. the groove in the double helix of DNA is occluded by the presence of nucleo-histone, the positively charged side chains of which are bound to the negatively charged phosphate groups in the backbone of the DNA molecule.

In B. there is an influx of positively charged ions into the nucleus under the influence of ecdysone which compete for and displace nucleohistone from stretches of DNA. This denuded DNA can now act as a template for the synthesis of RNA.

We can conclude that ecdysone initiates the events of moulting by depressing various genes which modify larval metabolism by the production of messenger RNA. It has been suggested (Kroeger 1967) that ecdysone produces its effects by altering the permeability and transport functions of the cell membrane, or the nuclear membrane, or on both. If a micro electrode is inserted into a chironomid salivary gland cell one can record an electrical potential across the cell membrane (page 22 for a general discussion of trans-membrane potential). This is presumably due to a gradient of distribution of ions, notably potassium (K^+) between the interior of the cell and surrounding tissue fluid. If ecdysone is added to the incubation medium containing isolated salivary gland this causes an increase in the recorded electrical potential. This change occurs one minute after addition of the hormone and lasts for about twelve minutes when the potential difference rises to a value of 15 m.v. higher than the previous value. Striking confirmatory evidence for this view of the mode of action of ecdysone is that the effect of ecdysone on puffing can be mimicked by altering the ionic environment of salivary gland cells. Some chromosome puffs appear after increases in the concentration of a range of ions, e.g. Na^+, or Li^+, or Mg^{2+}, or Ca^{2+}, or NH_4^+. Some puffs are more stringent in their requirements and develop only after increases in the concentration of one ion.

The mode of action of ions is at the moment a subject of speculation. In view of the rapidity of action of changes in ion concentration and because they are as effective on isolated nuclei as on whole cells (i.e. no cytoplasmic factor is necessary for their action) it has been suggested that the ions may act directly upon the chromatin, possibly on DNA—histone links (page 61). Thus ions could compete with histones for the negative charges on the DNA molecule and cause a displacement of histone revealing stretches of DNA which could now act as a template for the synthesis of RNA (Fig. 42).

CHAPTER 9

Growth Hormone: An Action on Ribosomal Translation of Messenger RNA

The discovery of messenger RNA and the publication of Jacob and Monod's concepts of the regulation of gene activity in baceteria were followed by enthusiastic application of these ideas to the problem of mechanisms of hormone action. We have seen how evidence has been obtained that some effects of hormones on protein synthesis are associated with the derepression of genes and the synthesis of new kinds of messenger RNA. In the larvae of lepidoptera these changes can be visualized on the giant polytene chromosomes.

It is, however, becoming clear that primary action at the gene level is by no means the only way in which hormones can alter the protein synthesis of the cell. Indeed there is evidence that messenger RNA is not necessarily a limiting factor in protein synthesis. In the egg of the sea urchin the act of fertilization results in a rapid stimulation of protein synthesis. But this occurs in spite of a complete absence of RNA synthesis. Messenger RNA must have been present in the egg before fertilization although this was not associated with protein synthesis. Other studies of protein synthesis in a liver tumour, showed that the rate of synthesis was very similar to normal liver tissue in spite of the fact that the tumour contained over four times as much RNA as in normal liver.

We shall, in this chapter, be considering some of the evidence that pituitary growth hormone exerts at least part of its effect on protein synthesis by modifying the ability of ribosomes to translate messenger RNA into terms of protein synthesis. Before looking at this effect of growth hormone we must first outline briefly what is known of the nature of this translation process on the ribosome.

Protein Synthesis

The Genetic Code. We have seen in Chapter 6 how the information for the construction of proteins is passed on from one generation to another in the DNA of the nuclear chromatin. This information is contained in the four-letter alphabet—*Adenine*, *Cytosine*, *Guanine* and *Thymine*—strung along the molecule of DNA. These four bases must, in some way, determine the sequence of the 20 amino acids of which proteins are composed. Each amino acid must be "coded for" by a group of bases. Obviously one base cannot "code" for one amino acid—there would only be four amino acids. Similarly two bases are not enough since only 16 different two-letter "words" can be made up from an alphabet of four letters (e.g. AA, CG, CC, CU and so on). Words of three letters (AAA, AGA, ACA, AUA, etc.) are thus the minimum. With three nucleotides per "word" the number of possible different combinations of A, C, G and T is 64. There are thus more "words"—codons—than there are amino acids to be coded for. In fact some amino acids are coded for by several different codons. Leucine, for example, can be coded for by six different codons.

Nonsense codons. Of the 64 possible codons, 61 have been shown to code for a particular amino acid. The three remaining codons UAG, UAA and UGA have been called nonsense codons. They do have, however, an important function in the genetic code in that they act as "full stops" on the molecule of messenger RNA at which point protein construction stops and the polypeptide is released.

Capital letter codons. These codons mark the beginning of "reading" of the message of RNA on the ribosome. In the bacterial E. Coli two codons seem to act as the "capital letters" for the code—AUG and GUG.

The Transcription Process. We have already seen (page 59) that the message of DNA is transcribed into a message carried by a special kind of RNA—messenger RNA—which leaves the nucleus to instruct protein synthesis on the ribosomes of the cell cytoplasm.

RNA is a single stranded molecule very similar to DNA except that one of the four bases is different (thymine of DNA is replaced by uracil) and the sugar in the backbone of the molecule is ribose rather than desoxyribose. It appears that only one of the DNA strands is copied in RNA synthesis—and since the two strands are different (although complementary) the same strand is copied on each occasion of RNA synthesis. In the copying, or transcribing process, G of DNA appears as C in RNA, C appears as G, A appears as U and T as A.

The Assembly of Amino Acids to Form Specific Proteins

Activating enzymes. The first stage in the joining of amino acids to produce protein molecules consists in a reaction of individual amino acids with ATP to form an adenyl-amino acid compound in which the energy of ATP is stored. This reaction is somewhat unusual in that when ATP activates a compound it usually does so by transferring the terminal phosphate bond to the compound. Here, however, the adenosine monophosphate combines with the amino acid and pyrophosphate is released.

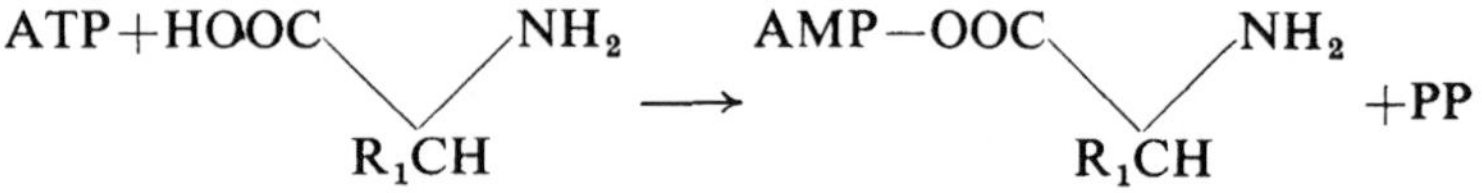

This reaction requires the presence of activating enzymes of which there are at least twenty different kinds. These enzymes catalyse this first important step in protein synthesis whereby the amino acid is charged with sufficient energy to unite with other amino acids by peptide links. These adenyl-amino acids, also called amino-acyl adenylates, are very reactive substances and if they are mixed together in a solution they will spontaneously react to form polypeptides (see below).

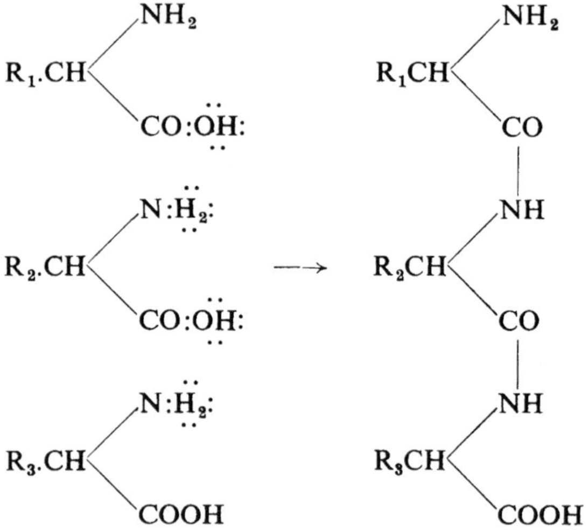

This random union of activated amino acids does not occur in living cells because each amino-acyl adenylate molecule remains bound to the activating enzyme which initiated its formation. The activating enzyme now performs a second function. It brings a molecule of transfer ribose nucleic acid (RNA) into contact with the amino-acyl adenylate.

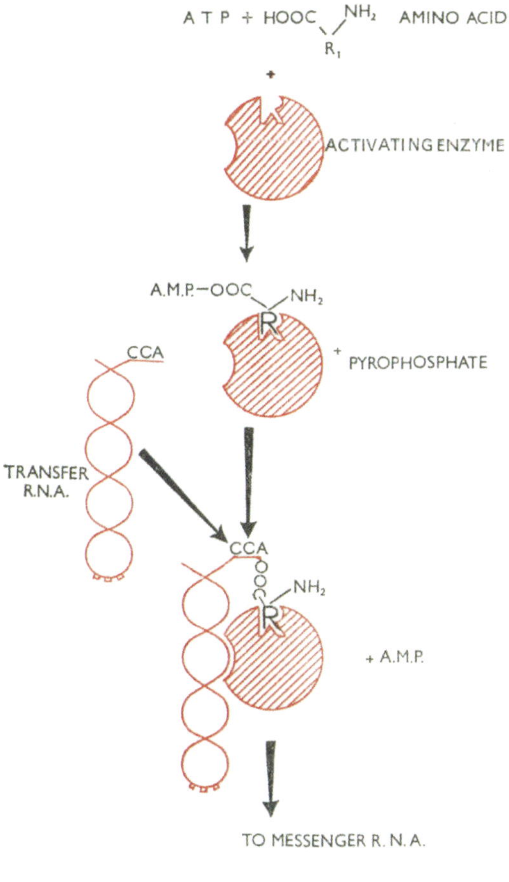

FIG. 43. *Stages in protein synthesis. 1. The activation of amino acids and attachment to transfer RNA.*

Two active sites are shown on the activating enzymes, one specific for the amino acid, the other for transfer RNA.

Transfer RNA. The structure of transfer RNA is such that in one region of the molecule it can combine with an amino acid and in another region it can combine with a special part of the molecule of messenger RNA which is the template on which protein is constructed. Compared to other nucleic acids it is a fairly small molecule containing 65–70 nucleotides. The molecule is a single chain which is bent back upon itself to form a double helix with a loop at one end (Fig. 43). In this terminal loop is the site for the attachment of amino acids and if this loop is removed from the molecule amino acids cannot become attached. However, since all types of transfer RNA's seem to have the same terminal base sequence in this loop (cytidylyl-ĉytidylyl-adenylic acid) there must be some other part of the molecule which recognizes individual amino acids. There are large numbers of different kinds of transfer RNA which are specific for each amino acid, sometimes for more than one. At the point on the molecule where the strand folds back on itself there are some unpaired bases which probably form the recognition site for the RNA template.

The Translation Process. The next stage in the synthesis of protein is the attachment of transfer RNA's with their attached amino acids to the messenger RNA template. Messenger RNA does not appear to operate as a template in protein synthesis until it has been bound to ribosome. Each molecule of messenger RNA becomes related to a group of ribosomes (to form a polysome) during protein synthesis. It is thought that the ribosomes may actually move along the messenger molecule, reading and translating each codon as it progresses along the molecule.

The exact way in which the complex of transfer RNA-activated amino acid is transferred from the activating enzyme to the molecule of messenger RNA has not been elucidated. It is thought that the transfer RNA molecules with their attached amino acids are assembled on the messenger RNA in order by means of the pairing of bases on the transfer RNA molecule with complementary bases on the messenger RNA. When this happens peptide bond formation occurs between adjacent amino acids and the protein peels off the ribosome. After releasing their amino acid the molecules of transfer RNA leave the ribosome (Fig. 44). The incorporation of amino acids into the growing peptide chain is thought to occur at the site of attachment of messenger RNA to the ribosome. This site changes of course as the ribosome moves

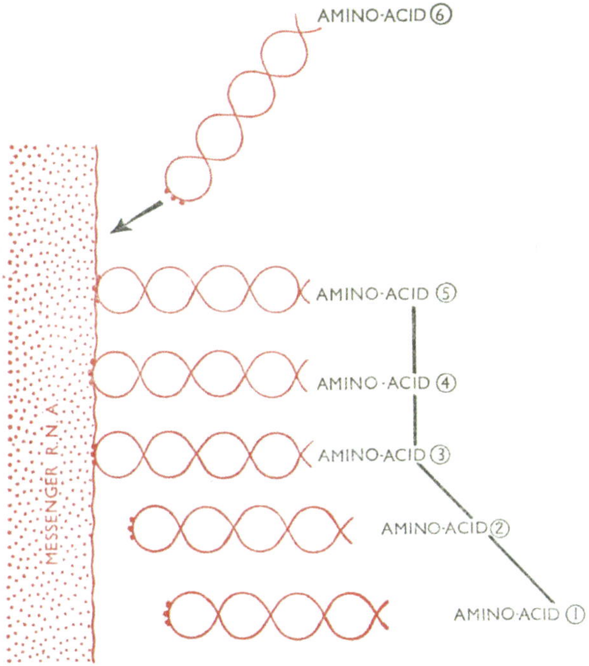

FIG. 44. *Stages in protein synthesis. 2. The assembly of the protein molecule on the surface of messenger RNA (ribosome not shown).* Molecules of transfer RNA bearing activated amino acids are shown becoming attached to messenger RNA where peptide bond formation occurs between adjacent amino acids. The developing peptide chain is shown peeling off the messenger RNA. The members of amino acids shows the sequence of amino acid incorporation.

---

along the messenger RNA molecule, translating the code of messenger RNA into amino acid sequence as it does so.

Growth Hormone

No attempt will be made here to give a complete account of growth hormone, its physiological effects, mode of action and the regulation of its secretion from the anterior pituitary gland. The physiological consequences of administering the hormone to animals are complex and involve not only widespread effects on metabolism of protein, carbohydrates and lipids but also inter-relations with other hormones such as insulin and glucocorticoid hormones. Accurate estimations of the quantity of the hormone in body fluids has only recently become possible. The true physiological significance of growth hormone is, for these various reasons, still a matter of controversy and speculation. It has indeed been doubted that the hormone has a direct effect upon growth, and it has been suggested that the growth-promoting effects of the hormone are due to its ability to stimulate the secretion of another hormone—insulin. The hormone is included in this volume because recent researches into the mechanism of its action provide us with an interesting example of regulation at the level of ribosomal translation of messages arising in the nucleus.

The Pituitary Gland and Growth. We can define growth as an increase in body weight and size due to the laying down of tissues which have a similar composition to those of the original body. This definition thus excludes changes in body weight and size due to accumulation of water or fat. More precise indices of growth of an animal are increases in the amount of body protein or the size of the bones.

The significance of the pituitary gland for growth was first indicated by observations of a rare disease of man first described in 1886, in which "overgrowth" of various tissues of the adult (e.g. bones, skin, tongue, abdominal organs) are associated with a particular tumour of the anterior pituitary gland. This condition is called acromegaly. In acromegaly there is little extra growth of long bones (femur, tibia, etc.) because these bones grow in length only at the disc of cartilage which separates the shaft from the head of the bone—the epiphysis. This cartilaginous disc is transformed into bone during adolescence. If over-activity of the anterior lobe of the pituitary gland occurs before this transformation has occurred then marked growth of long bones becomes possible—with the production of a giant. The reverse condition of dwarfism results from deficiencies of pituitary growth hormone.

The pituitary gland influences growth in various ways. First, by regulating other endocrine glands—ovary, testis, thyroid—the secretions of which influence growth. Second, by the production of a specific growth hormone, also called somatotrophin. This hormone is a protein whose exact chemical structure awaits elucidation.

Some Effects of Administration of Growth Hormone. In animals in which the epiphyses of the long bones are still unfused to the shaft the administration of growth hormone stimulates the growth of both bone and soft tissues to produce a symmetrical enlargement of the animal. Treatment of rats with growth hormone can produce individuals which are twice the size of normal individuals, such is the potency of the hormone. Nearly all organs and tissues participate in the response to growth hormone. Biochemical studies show that this growth is due to an increased synthesis of proteins from amino acids. At the cellular level growth hormone stimulates the synthesis of RNA of all types, nuclear RNA, transfer RNA, Messenger RNA and ribosomal RNA.

The effects of removal of the anterior pituitary gland are the reverse of the above changes. In the growing animal there is a virtual cessation of growth. In adults there is a loss of body weight. Tissues removed from these animals show depression of the incorporation of amino acids into protein, a fall in the RNA content of cells and a decline in the number of ribosomes of the cell. The administration of purified growth hormone can reverse these changes.

The Mechanism of Action of Growth Hormone on Protein Synthesis. The biochemical effects of growth hormone, as described above, would be consistent with the idea that growth hormone acts at the genetic level by promoting increased synthesis of messenger RNA ultimately leading to increases in protein synthesis. The rate of RNA synthesis can be studied in isolated tissue preparations by studying the rate of incorporation of radioactively labelled RNA precussors into RNA. These studies showed that the administration of growth hormone stimulated the labelling of RNA. Studies of this labelled RNA showed that some of it had the characters of messenger RNA. These effects are similar to those produced by hormones such as ecdysone, oestradiol and cortisone.

The effects of removal of the anterior pituitary gland provided further support for an action of the hormone on chromatin. This procedure in the rat results in a decrease in the number of polysomes (page 94) in the liver which suggests that removal of growth hormone had reduced RNA production and hence formation of polysomes.

This view of the action of growth hormone was dealt a blow by the discovery that although the effect of growth hormone on the synthesis of RNA could be blocked by actinomycin (page 63) this did not prevent the stimulation of protein synthesis by ribosomes of liver or diaphragm. We have already seen (page 90) that the amount of messenger RNA in cells is not always a limiting factor in protein synthesis. Further support for this is provided by experiments studying the effects of removal of the anterior pituitary gland on the rat heart. This procedure reduced the rate of protein synthesis of heart muscle although the number of polysomes (i.e. combinations of messenger RNA and ribosomes) was not altered.

It would seem that growth hormone can stimulate protein synthesis in a way which does not necessarily involve an increased rate of production of messenger RNA. Professor Korner has provided evidence that a further site of action of growth hormone in the cell is the ribosome. The removal of the pituitary gland reduces the ability of ribosomes to assemble amino acids into proteins and this effect can be reversed to a degree by treatment with growth hormone. This change in the activity of the ribosomes seems to be associated with a reduced ability of the ribosomes to attach themselves to messenger RNA.

We can conclude that the action of growth hormone on protein synthesis is not readily explainable by "switching on" of particular genes. Certainly increases in RNA synthesis do occur after treatment with growth hormone but this may well be secondary to actions of the hormone in the cytoplasm, in particular an increased association of messenger RNA and ribosomes. Obviously further information is needed to unravel this problem of growth hormone action.

Index

NOTES

NOTES

NOTES

NOTES

NOTES

Printed in Dunstable, United Kingdom

91221468R00065